Coatings Tribology

Coatings Tribology

Editors

Braham Prakash
Jens Hardell

MDPI • Basel • Beijing • Wuhan • Barcelona • Belgrade • Manchester • Tokyo • Cluj • Tianjin

Editors
Braham Prakash
Division of Machine Elements,
Luleå University of Technology
Sweden

Jens Hardell
Division of Machine Elements,
Luleå University of Technology
Sweden

Editorial Office
MDPI
St. Alban-Anlage 66
4052 Basel, Switzerland

This is a reprint of articles from the Special Issue published online in the open access journal *Coatings* (ISSN 2079-6412) (available at: https://www.mdpi.com/journal/coatings/special_issues/coat_tribol).

For citation purposes, cite each article independently as indicated on the article page online and as indicated below:

LastName, A.A.; LastName, B.B.; LastName, C.C. Article Title. *Journal Name* **Year**, *Article Number*, Page Range.

ISBN 978-3-03943-693-4 (Hbk)
ISBN 978-3-03943-694-1 (PDF)

Cover image courtesy of Pouria Valizadeh Moghaddam.
TiAlN PVD coating deposited on WC/Co substrate showing the transfer of 316L stainless steel after tribological test.

Contents

About the Editors

Braham Prakash obtained his BSc Engineering Mechanical degree (1974) from Punjab Engineering College Chandigarh and his MTech (Mechanical Engineering) and PhD (Tribology) (1976, 1993) from the Indian Institute of Technology Delhi (India).

He is presently Professor Emeritus at Luleå University of Technology (Sweden) and a Distinguished Visiting Professor at the Tsinghua University (China). Earlier, he was a Professor and Head of Tribolab at the Division of Machine Elements of Luleå University of Technology (2002–2019), a faculty at Indian Institute of Technology Delhi (1981–2002) and R&D professional in industry (1976–1981).

He was a Visiting Researcher at Tokyo Institute of Technology (1985) and Fellow of Japan Society for the Promotion of Science (JSPS) at Chiba Institute of Technology (1998–2000). He was Visiting Professor at Tokyo University of Science (2016) as well as at Indian Institute of Technology, Ropar (2010–2013).

His research and teaching activities pertain to high temperature tribology; tribology of materials and lubricants; solid lubricants/self-lubricating coatings; boundary lubrication; tribology of machine components (bearings, gears and seals); analysis of wear problems; and tribotesting.

Jens Hardell obtained his MSc degree in Mechanical Engineering in 2005 from Luleå University of Technology. In 2009, he completed his PhD in Machine Elements at Luleå University of Technology. He became Associate Professor at the Division of Machine Elements, Luleå University of Technology, in 2014, as well as Head of Division in 2016.

He was also a visiting Professor at Université de Lyon, Ecole Nationale d'Ingénieurs de Saint-Etienne in France (2017).

His main research and teaching interests includes high temperature tribology, friction and wear in dry contacts, tribomaterials, surface engineering for friction and wear control, and wear and failure analysis.

Preface to "Coatings Tribology"

Tribological coatings are increasingly used to control (mostly to minimize) friction, wear, and surface damage to move machine components in various applications. The use of coatings, in addition to improving the efficiency and durability of tribological systems, also enables conserving strategic materials and minimize or eliminating the use of hazardous materials. The last few decades have seen massive developments in the field of coatings tribology. A Special Issue of Coatings (ISSN 2079-6412) dealing with various aspects pertaining to coatings tribology was published in 2018. This reprint book is based on this Special Issue and contains 11 research papers dealing with plasma sprayed, ion-beam assisted and sputtered coatings of different compositions. Some of the salient coatings include: NiCr-TiB2 and AlCoCrFeNiTi/Ni60 (plasma sprayed); Cr-Ti.B-N, stainless steel-silver, VCN-Cu, Cu-Al/MoS2, DLC/MoS2 (magnetron sputtered); Ag-MoCo (ion-beam-assisted bombardment); multilayered AlCrN (cathode arc PVD) and polymer composite coatings. The coatings reported in these papers have been deposited on different substrates (including soft substrates) for controlling friction and wear under different operating conditions. The characterization and application aspects of coatings have also been discussed in some of these contributions.

It is hoped that this book will be useful to all those engaged in the control of friction and wear in diverse tribological applications and provide an impetus for further research on this important topic.

The editors would like to take this opportunity to thank MDPI, in particular Ms. Zhiqiao Dong and Ms. Flora Ao for their initiative and support in first bringing out the Special Issue and then this reprint book on coatings tribology.

Braham Prakash, Jens Hardell
Editors

Article

Composition versus Wear Behaviour of Air Plasma Sprayed NiCr–TiB_2–ZrB_2 Composite Coating

Ning Zhang [1], Nannan Zhang [1,*], Sheng Guan [2], Shumei Li [2], Guangwei Zhang [2] and Yue Zhang [1]

1 Department of Material Science and Engineering, Shenyang University of Technology, Shenyang 110870, China; zhangning201610034@sut.edu.cn (N.Z.); yuezhang@sut.edu.cn (Y.Z.)

2 Dalian Huarui Heavy Industrial Special Spare Parts Co., Ltd., Dalian 116052, China; Guansheng@dhidcw.com (S.G.); LiSM@dhidcw.com (S.L.); lichao@dhidcw.com (G.Z.)

* Correspondence: zhangnn@sut.edu.cn; Tel.: +86-242-549-6812

Received: 13 June 2018; Accepted: 5 August 2018; Published: 6 August 2018

Abstract: The NiCr–TiB_2–ZrB_2 composite coating was deposited on the surface of blades made of steel (SUS304) using high-energy ball milling technology and air plasma spraying technology, which aimed to relieve the wear of the blades during operation. The influence of titanium diboride (TiB_2) and zirconium diboride (ZrB_2) on the microstructure and wear resistance of the coatings was investigated by X-ray diffraction, scanning electron microscopy, Vickers microhardness tester, and a wear tester. The results showed that the TiB_2 and ZrB_2 particles were unevenly distributed in the coatings and significantly increased the hardness and anti-wear, which contributed to their ultra-high hardness and extremely strong ability to resist deformation. The performance of the coatings was improved with the increase of the number of ceramic phases, while the hardness and wear resistance of the coating could reach their highest value when the TiB_2 and ZrB_2 respectively took up 15 wt.% of the total mass of the powder.

Keywords: TiB_2; ZrB_2; coating blade; anti-wear

1. Introduction

The coating of a blade has a crucial influence on the quality of fine art coated paper and coated white paper. However, with the further development of coating technology resulting in a higher production efficiency, the speed of the coating blade has also been increased from 5 to 16 m/s. The traditional carbon steel blade is gradually being replaced by a wear-resistant ceramic coating blade because its service life is too short, it is about 3–4 h, as a result of severe three-body abrasive wear. The process of changing blades and the running-in stage has reduced the efficiency of the coating production [1].

At present, countries around the world are using the oxide ($Al_2O_3 \cdot TiO_2$ and ZrO_2) and carbide (Cr_3C_2 and WC) ceramic blades produced by the three companies of BTG (Eclépens, Switzerland), GREEN COAT (Siheung, Korea), and PSK (Portland, OR, USA) [2]. The performance of oxide ceramic blades has been significantly improved compared to the carbon steel blades, as the main component of the coating is kaolin, which contains $Al_2O_3 \cdot 2SiO_2 \cdot 2H_2O$, $CaCO_3$, TiO_2, and so on. Furthermore, the intersolubility of the same materials is large. However, adhesion wear can easily occur between these oxide ceramics [3], and thus, the oxide ceramic coating is not very suitable for practical production applications. Tungsten carbide (WC) has a relatively high thermal conductivity [4] and the thermal conductivity rose linearly from 29.3 to 86 W/(m·K) when the temperature increased from 20 to 500 °C [5]. Therefore, it is easy to transfer a large amount of heat to the blade steel strip during the initial stage of contact between the blade and the paper, causing the blade edge to become uneven, which is unfavorable for the production of coatings. The hardness of chromium carbide (Cr_3C_2) is relatively low, and its wear resistance is far inferior to the WC [6–8]. Therefore, a new type of ceramic coating is urgently needed to solve these problems.

Boride ceramics (TiB_2 and ZrB_2) have a higher hardness (Moh's hardness of 9–10) and an outstanding wear resistance. Compared to WC (15.63 g/cm^3), boride ceramics have a density that is lower by about 75% (4.51 and 5.8 g/cm^3), and a lower thermal conductivity (24.3–23.03 W/(m·K) and 24.28–41.87 W/(m·K) in the temperature range of 20–500 °C [9–11]. Therefore, boride ceramics are potential materials for replacing the oxides and carbide ceramics.

Jones et al. [12] prepared Fe(Cr)–TiB_2 coatings using self-propagation high-temperature synthesis (SHS) and high velocity oxygen-fuel spraying (HVOF) technology. They found that the wear-coefficient of a 30% FeCr–TiB_2 coating (7.235 μg/(m·N)) was lower than that of a 17% Co–WC (10.707 μg/(m·N)), 30% Fe(Cr)–TiC (9.061 μg/(m·N)), and 25% Cr_3C_2–NiCr (19.069 μg/(m·N)) coating. In addition, the wear rate of the TiB_2-based composite coating was only 50% of the TiC-based coating, and was far superior to other materials when the grinding material was abrasive silica. Xu et al. [13] used SHS and atmospheric plasma spraying technology to prepare ZrC–ZrB_2/Ni composite coatings with a variety of different proportions of Ni. The study showed that the highest hardness of coating reached 525 ± 96 $HV_{0.1}$ and had the maximum resistance to wear when the content of Ni was 10%.

The thermal conductivity of ZrB_2 is lower than that of the TiB_2 coating, while the resistance to wear of TiB_2 is better than that of the ZrB_2 coating [14]. Most importantly, the addition of TiB_2 and ZrB_2 in the same coating can significantly improve its wear resistance [15–17]. Hence, the TiB_2–ZrB_2 ceramic composite coatings were the best choice to improve the quality of the coating blade. However, the pure ceramic is too brittle, so it is necessary to add a ductile metal (NiCr) with a good wettability to form a cermet composite material in order to better meet the use conditions [18,19]. Furthermore, plasma spraying is especially suitable for ceramic materials with a high melting point because of the higher flame temperature and faster jet speed of the plasma spraying. In this paper, the air plasma spraying technology was used to prepare a NiCr–TiB_2–ZrB_2 composite coating, and its tribological properties were analyzed.

2. Materials and Methods

2.1. Feedstock Materials

Commercially available nickel chromium (NiCr), TiB_2, ZrB_2 powders, and SUS304 stainless steel were purchased in this study. The NiCr powder (Ti Metal Materials Co., Ltd., Changsha, China) was atomized into a spherical shape by an inert gas with an average size of 50 μm. The TiB_2 and ZrB_2 powders were obtained from Xiangtian Nano Materials Co., Ltd. (Shanghai, China) as irregular powders with an average particle size of 50 μm. The substrates (SUS304) were provided by Shenzhen Bao Metal Products Co., Ltd. (Shenzhen, China), which have dimensions of approximately 30 mm × 50 mm × 5 mm. The information on all of the feedstock materials is listed in Table 1.

Table 1. Composition of the raw materials (at.%).

Materials	Purity	Zr	Ti	B	Cr	Ni	Fe	C	Others
NiCr	99.0	–	–	–	19.86	80.08	–	0.01	Bal.
TiB_2	99.5	–	68.23	30.75	–	–	0.14	0.13	Bal.
ZrB_2	99.9	79.27	–	19.30	–	–	0.09	0.15	Bal.
SUS304	–	–	–	–	18.65	9.73	68.54	0.07	Bal.

2.2. Powder Preparation

The first stage of the experiment involved the preparation of NiCr–TiB_2–ZrB_2 powders by the high-energy ball milling method. Firstly, we mixed the three types of powders in equal proportions, before the mixture of the powders and the same quality of agate balls (ball-to-powder weight ratio of 1:1) were placed in the same agate jar with a high purity argon (99.999%) protection. This was then placed in the ball mill (XQM-0.4, Delixi Hangzhou Inverter Co., Ltd., Hangzhou, China) at 6.67 r/s for 640 min. The mixture was rested for 20 min every 80 min to maintain the powders at a lower temperature. After this, the moisture was removed from the powders in a digital electric blast dry oven (101A-2, Shanghai JinPing Instrument Co., Ltd., Shanghai, China). Finally, three different types of target cermet composite powders (NiCr (5 wt.%)–TiB_2 (5 wt.%)–ZrB_2, NiCr (10 wt.%)–TiB_2 (10 wt.%)–ZrB_2, and NiCr (15 wt.%)–TiB_2 (15 wt.%)–ZrB_2) were prepared by adjusting the proportions of NiCr, TiB_2, and ZrB_2.

2.3. APS Spraying Experiments

The NiCr–TiB_2–ZrB_2 composite coatings were prepared on the surface of the SUS304 using the SG-100DC plasma torch (Praxair 3710, Cleveland, OH, USA) and ball milled powders. The parameters of the spraying are summarized in Table 2. Prior to the spraying, one side of the substrate was grit-blasted with 0.6 mm of Al_2O_3 particles to achieve the required experimental surface. The Ar served both as a protective gas and as a powder-feeding gas to reduce the degree of oxidation, while H_2 was used as a fuel gas to provide energy for the experiment during the process of spraying. The specimens were cut into small pieces to facilitate the investigation of the morphology and phase composition after the experiments.

Table 2. Operation parameters of plasma spraying.

Parameters	Value
Current (A)	600
Voltage (V)	40
Ar (dm^3/s)	0.67
H_2 (dm^3/s)	0.17
Powder feed rate (g/s)	0.58–0.75
Gun traverse speed (mm/s)	90
Spray distance (mm)	100
Pre-heating temperature (°C)	120

2.4. Microstructural Characterisation

A scanning electron microscope (SEM, S-3400, Hitachi, Tokyo, Japan) with an energy dispersive spectrometer (EDS) ($V_{a/c}$ = 20.0 kV) was employed to observe the microstructure of the cross-section and the worn surface topographies, as well as for the local elemental analysis. The X-ray diffraction equipment (XRD, Shimadzu 7000, Shimadzu, Kyoto, Japan) using monochromatic Cu Kα radiation (λ = 0.1541 nm) at 40 kV, 30 mA was used to distinguish the phases composition of the powders and coatings, which had a scanning range of 20°–90° and a scanning speed of 4°/min.

2.5. Performance Tests of Coatings

In order to reduce the roughness to less than 0.01 μm and to present mirror conditions, which can minimize the influence of surface roughness on the hardness and tribological properties of the coatings as much as possible, it is necessary to polish the cross section and upper surface of the specimens before the performance tests. After this, the microhardness of the coatings was measured by means of an HVS-1000 Vickers microhardness tester (Jiangdong Kunning, Ningbo, China) with a 0.1-kg (0.98 N) load and 15-s dwell time. The average microhardness value was calculated on the base of ten measurements. In addition, the distance between every two indentations in this experiment is more than three imprints, in order to ensure that the individual indentations do not interfere with each other. The Al_2O_3 balls (with a diameter of 6 mm) were used as a counterpart to better simulate the actual application environment of the coating blades. The ability of the coatings to resist wear was tested through the ball-on-disk wear test (SFT-2M, Zhongke Kaihua Technology Development Co., Lanzhou, China) with a 30-N loading at a sliding speed of 0.4 m/s for 30 min, so the reciprocating wear distance is about 720 m. The experiments were performed in triplicate on the same sample in order to reduce the random error. After the wear test, the three-dimensional morphology of wear scratches was observed through laser scanning confocal microscopy (OLS4100, Olympus, Tokyo, Japan).

3. Results and Discussion

3.1. Microstructure and Phase Composition

The XRD analysis results for the NiCr–TiB_2–ZrB_2 powders are shown in Figure 1. It can be seen that the strongest peaks are attributed to Ni, while the remaining small peaks are referred to TiB_2 and ZrB_2 as well as the newly generated CrB phase. This meant that the powders obtained energy during the ball milling process, which led to the decomposition of the old phases (NiCr, TiB_2, and ZrB_2) and the generation of the new phase (CrB). In addition, there were no common oxidation phases [20], such as TiO_2, Ti_2O_3, $NiTiO_3$, and ZrO_2 that reduced the performance of the coating in the powders prepared using the high-energy ball milling method, which illustrates that the quality of the powders is relatively good.

The XRD diagrams of the three powders are almost the same, except for the different intensity of the peaks. This may be because the composition of the three powders is consistent, but the proportions of each powder are different, which leads to a difference in the content of the reactants, even under the same ball milling parameters. It can also be seen from Figure 1 that with an increase in the TiB_2 and ZrB_2 content, the corresponding peak intensity gradually increases, but the change in CrB is relatively small. The mechanism of the content of the generated materials needs further study.

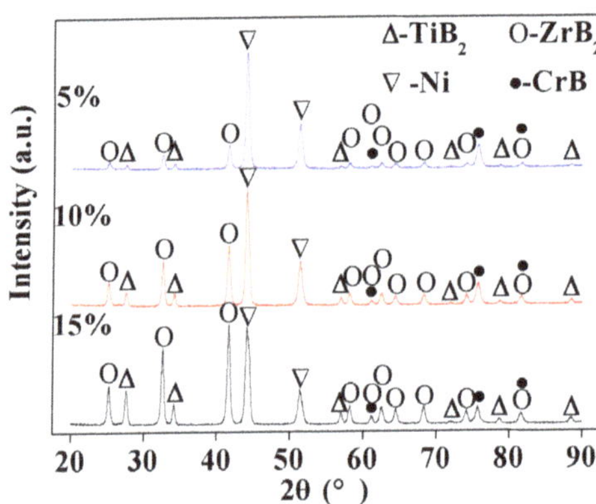

Figure 1. X-ray diffraction (XRD) spectra of three kinds of NiCr–TiB_2–ZrB_2 powders.

The XRD patterns obtained from the top surface of the as-deposited coatings are shown in Figure 2. It was found that the ceramic reinforcement phases of TiB_2 and ZrB_2 still exist in the coatings, although the peak intensities are significantly lower compared with the mixed powders. This is a good indication that the TiB_2 and ZrB_2 phases in the coatings are the un-reacted reactants from the starting powders, not the new generators. It is difficult to decompose the TiB_2 and ZrB_2 particles during the rapid melting and solidification process, so they are used to prepare the ceramic reinforced base coatings in this experiment. In addition, there are also some minor peaks existing in the coatings, such as BNi_2 and Ni_4B_3. These minor peaks can be attributed to the liberated particles formed by a small amount of decomposition of the TiB_2 and ZrB_2 particles that chemically react with NiCr.

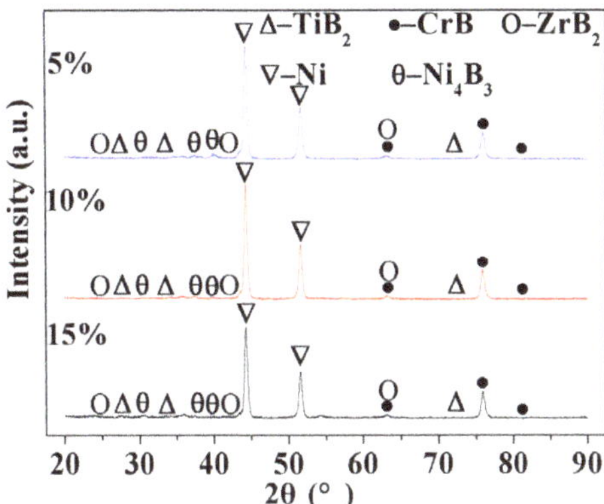

Figure 2. X-ray diffraction (XRD) spectra of three kinds of NiCr–TiB_2–ZrB_2 coatings.

Figure 3 shows the upper surface morphology of the NiCr–15ZrB_2–15ZrB_2 composite coatings with the distribution of all of the elements. The process of the mechanical stacking of plasma spraying powders is clearly shown in Figure 3a, and the pores of the coating were small, which proved that the density of the coating was relatively good. In addition, ZrB_2, TiB_2, and the compounds of the Ni, B, and Cr existing in the coating can be analyzed from Figure 3a–f and the results of the XRD.

The cross-sectional morphology of the NiCr–TiB_2–ZrB_2 composite coating using different proportions of source materials in Figure 4, showed that all of the coatings had a dense layered structure [21]. This is due to the fact that individual particles are rapidly heated and accelerated by Ar and H_2 in air plasma spraying. Typically, these particles are partially melted before being deposited in the substrate, because the temperature is already close to the melting point of the materials (3000 °C). After this, they will quickly cool to a lower temperature and flatten to form a thin sheet when they hit the substrate.

As observed from Figure 4a,c,e, the thicknesses of the composite coatings are in the range of 150–200 μm. All of the coatings have a small amount of porosity, especially in the interface between the coating and the substrate. This may be because the temperature difference between the coating and substrate is large, so the deformations of the particles have a significantly condensation contraction in this area, which finally produce pores. Furthermore, pores are also formed when there is not a complete covering between the different layers during the process of the coating deposition [22].

As seen from Figure 4b,d,f, the porosity of the coatings increases with increase of the ceramic phases. This may be due to the spraying process, as the degree of rebound of the ceramic phases is much greater than that of the metallic binder phases. Thus, with an increase in the ceramic phases, the gaps in the coating process increase gradually and the degree of melting of the ceramic phases is considerably less than the metal bonding phase. This reduces the probability that these gaps will be filled in time, resulting in an increased porosity. In addition, although this reduces the porosity of the three types of coatings, increasing the content of the bonding phase can significantly reduce the probability of the ceramic phase rebound, which subsequently reduces the porosity [23,24].

Based on the microstructure of Figure 4 and the energy spectrum data in Table 3, the phase composition of the coating can be simply divided into three categories. Firstly, the areas colored in black (A, F, and I in Figure 4) contain the pure TiB_2 that exists alone in the coatings. They are surrounded by large areas of grey bonding phases (Ni, CrB, BNi_2, and Ni_4B_3), as shown in B, E, and H in Figure 4. There are also striped mixed phases (C, D, and G in Figure 4), which contain bonding phases and ZrB_2. As the color of the stripe becomes darker, the concentration of ZrB_2 increases gradually. In these coatings, the wear resistance is mainly provided by the TiB_2 and ZrB_2 phases, while the bonding phases mainly increase the toughness and reduce the difference of the coefficients of expansion between the substrate and the coating. Therefore, a reasonable proportion of ceramic/metal can be conducive to improve the comprehensive performance of the coating.

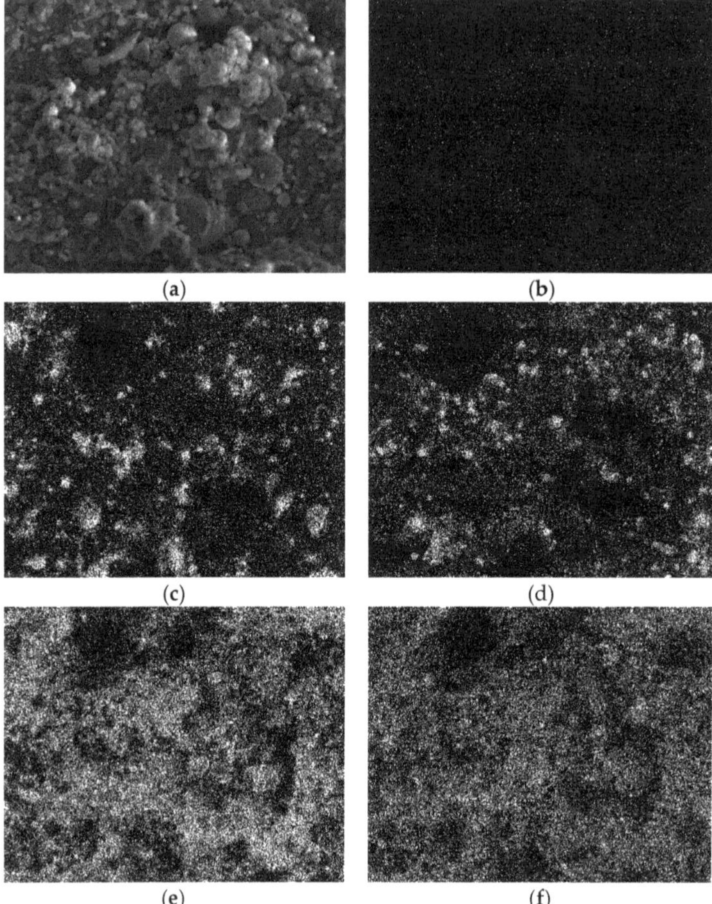

Figure 3. Elements distribution on the upper surface of NiCr–TiB_2–ZrB_2 coating: (**a**) NiCr–TiB_2–ZrB_2 coating; (**b**) B; (**c**) Ti; (**d**) Zr; (**e**) Ni; and (**f**) Cr.

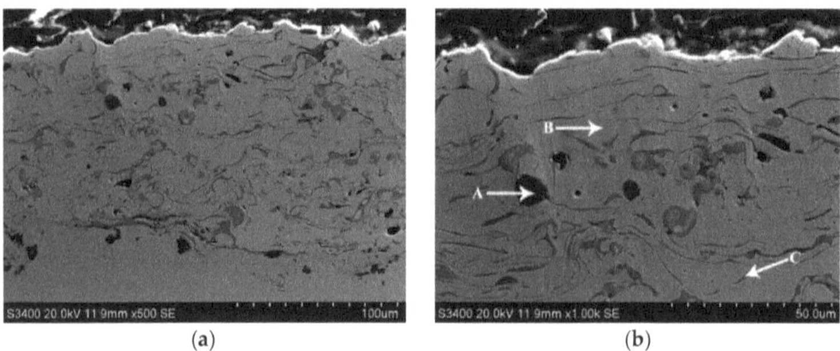

Figure 4. *Cont.*

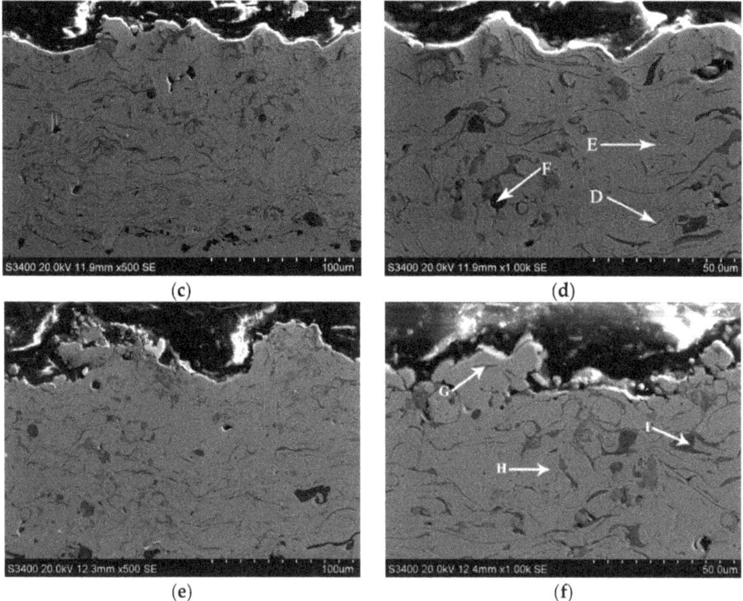

(c) (d) (e) (f)

Figure 4. Microstructure of the coatings: (**a**) $5TiB_2$–$5ZrB_2$–NiCr; (**b**) higher magnification image of $5TiB_2$–$5ZrB_2$–NiCr; (**c**) $10TiB_2$–$10ZrB_2$–NiCr; (**d**) higher magnification image of $10TiB_2$–$10ZrB_2$–NiCr; (**e**) $15TiB_2$–$15ZrB_2$–NiCr; and (**f**) higher magnification image of $15TiB_2$–$15ZrB_2$–NiCr.

Table 3. Chemical composition (at.%) in different regions in Figure 4.

Zone	Ni	Cr	Zr	Ti	B
A	–	–	–	34.21	65.79
B	72.43	15.98	0.24	0.62	10.73
C	15.93	19.98	10.07	12.41	41.61
D	29.95	14.95	10.71	7.82	36.57
E	61.13	17.64	0.62	1.25	19.36
F	–	–	–	36.47	63.53
G	20.68	12.06	5.73	13.70	47.83
H	65.43	14.91	0.11	0.83	18.72
I	–	–	–	33.52	66.48

3.2. Performance Test of Coatings

It can be seen from Figure 5 that the average hardness obtained on the coating cross-section was 753, 957, and 1102 $HV_{0.1}$, respectively. However, the microhardness value of each coating was floating up and down in a range. This is due to several reasons. Firstly, the parts of the ceramic phases do not dissolve in the bonding phases, which prevents uniform hardness. Furthermore, the presence of fine pores will reduce the hardness. In addition, it is attributed to the high hardness of the ceramic phases, as an increase in the proportion of ceramic results in a gradual increase in the hardness of the composite coating. Moreover, the generally accepted opinion in the field of friction and wear is that the ceramic phases in the metal ceramic composite coating are mainly able to provide wear resistance to the coating because of their high hardness. The main role of the metal phase is to maximize the retention of the ceramic phases in the coating and to improve the ductility of the coating and the bonding strength between the substrate materials due to the high toughness and the coefficient of thermal expansion being similar to the substrate. Therefore, a higher content of ceramic phases results in a greater hardness and better wear resistance when the toughness and bonding strength fulfill the requirements.

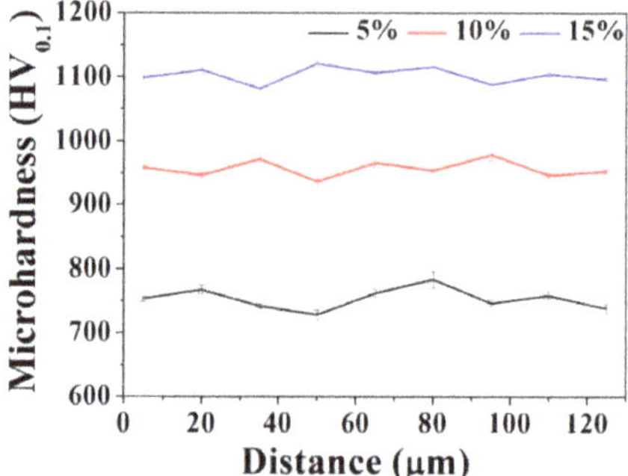

Figure 5. Microhardness value of the coatings.

The Weibull distribution is widely used in reliability engineering as we can easily infer its distribution parameters using probability values. The coating contains two types of phases (hard phase and binder phase) with a large difference in hardness, and the hardness measurement in the experiment is random, so the Weibull distribution can be used in this paper to analyze the continuity of the coating hardness from the view of probability and statistics. The formula can be described as follows [25]:

$$F(H) = 1 - \exp[-(H/\eta)^{\beta}] \tag{1}$$

where H is the hardness data, β is the Weibull shape parameter, and η is a normalized tested parameter. The equation can also be transformed into the following:

$$\ln[-\ln(-F(H))] = \beta[\ln(H) - \ln(\eta)] \tag{2}$$

As shown in Figure 6, all of the points can be synthesized into a straight line and the equations ($y = \beta x + b$) and reliability parameter (R) are listed separately in each diagram. With the addition of the ceramic phases (TiB_2 and ZrB_2), the β values of the coatings rose from 67.39 to 134.28, implying that the coating had a minimal variation when the content of the ceramic phases (TiB_2 and ZrB_2) in the powder reached 30% [26]. However, the range of the R values was 0 to 1, and an increase in the R value resulted in an increase in the coating stability. In addition, the R value of all of the coatings is greater than 0.9 and it can be seen that the overall change of the coatings was small.

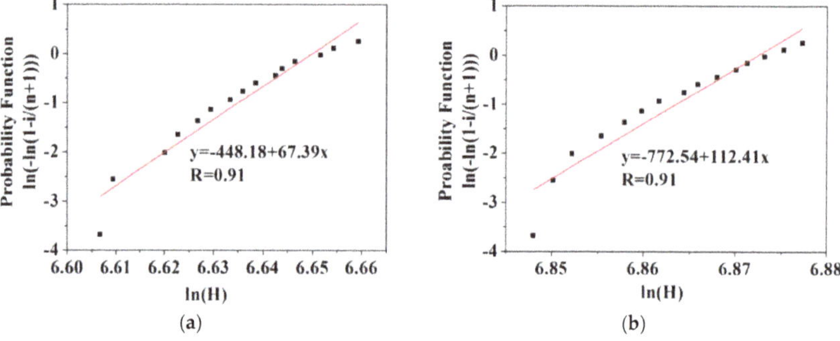

Figure 6. *Cont.*

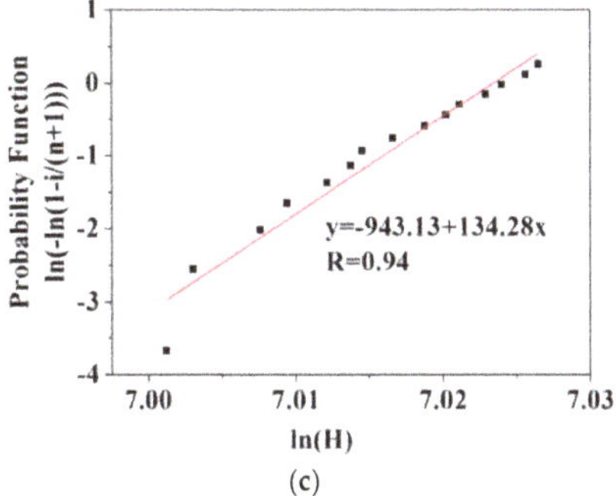

Figure 6. Linear fit of Weibull distribution of each load: (**a**) NiCr–5TiB_2–5ZrB_2; (**b**) NiCr–10TiB_2–10ZrB_2; and (**c**) NiCr–15TiB_2–15ZrB_2.

It can be seen from Figure 7 that the average friction coefficient curves of the three types of NiCr–TiB_2–ZrB_2 composite coatings against the Al_2O_3 ball were approximately 0.57, 0.50, and 0.45 in the steady wear stage, respectively. The friction coefficient is lower, with a large fluctuation found in the running-in stage, which may be attributed to the changes in the surface roughness. In general, the coefficient of friction is mainly related to the test parameters, the properties of materials, surface roughness, and lubrication state between the sample and grinding material (Al_2O_3). Considering that the three types of experiments differ only in the material composition, it can be understood that increasing the number of ceramic phases can reduce the friction coefficient of the coatings, which indicates that the ceramic phases are appropriate reinforcements to further enhance the wear resistance of the composite coatings. Moreover, as seen from the curves in Figure 6, the friction coefficients become more stable with an increase in the content of the ceramic phases, especially in the content of the ceramic phases TiB_2 and ZrB_2, which increased from 20 to 30 wt.%.

Figures 8–10 separately show the average wear depth and mass loss of NiCr–TiB_2–ZrB_2 composite coatings after the wear test, while the surface topographies corresponding to the mapping in Figure 9 are inserted in the top-left. They show a trend that is similar to the friction coefficient of the NiCr–TiB_2–ZrB_2 (Figure 6), which gradually decreases with the increased content of the ceramic phases, although the decreasing trend was less marked. In addition, the NiCr–15TiB_2–15ZrB_2 had a minimal wear loss (0.00784 g) and the shallowest wear depth (19 μm) of about 39.17% and 60.31% of the maximum, respectively, which showed the best anti-wear performance among the three types of coating.

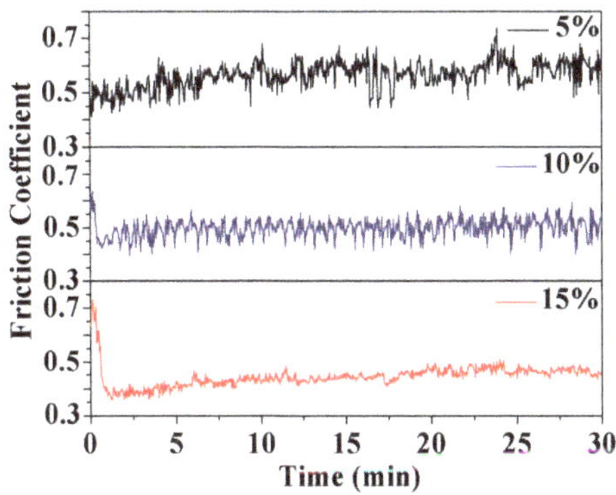

Figure 7. Friction coefficient of the NiCr–TiB_2–ZrB_2 coatings.

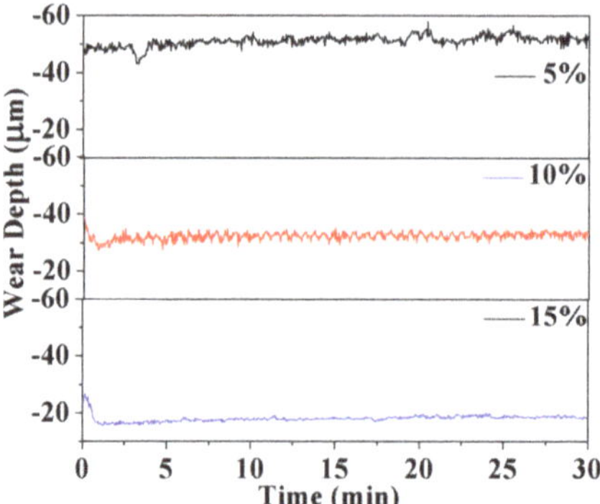

Figure 8. Wear depth of NiCr–TiB_2–ZrB_2 coatings.

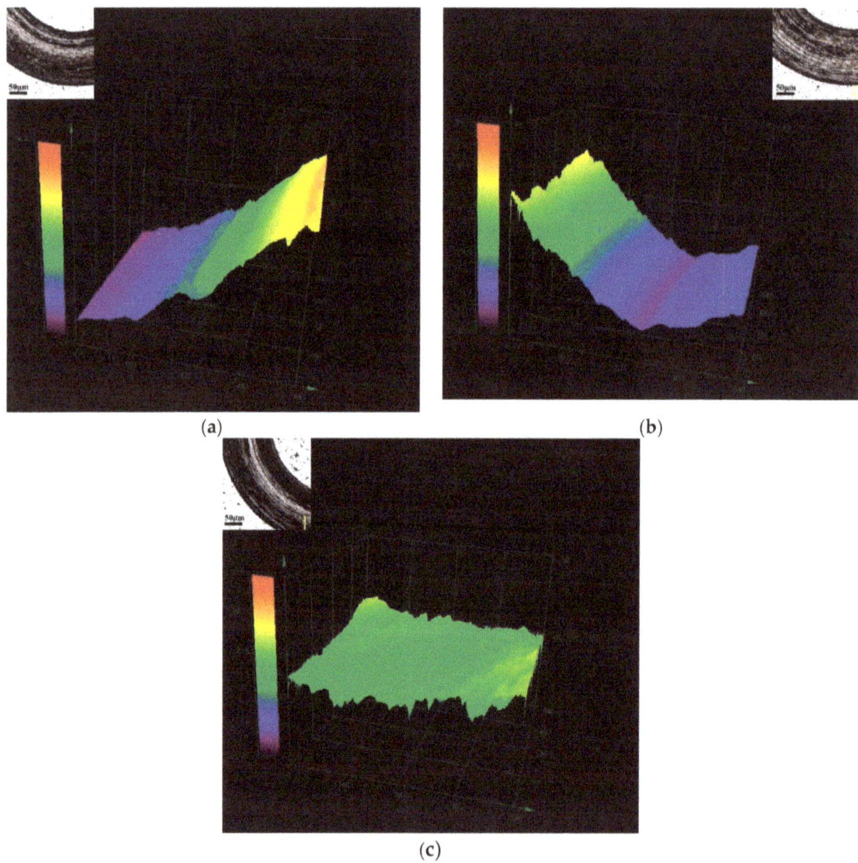

Figure 9. Three-dimensional (3D) non-contact surface mapping of the wear scars of (**a**) NiCr–5TiB_2–5ZrB_2; (**b**) NiCr–10TiB_2–10ZrB_2; and (**c**) NiCr–15TiB_2–15ZrB_2.

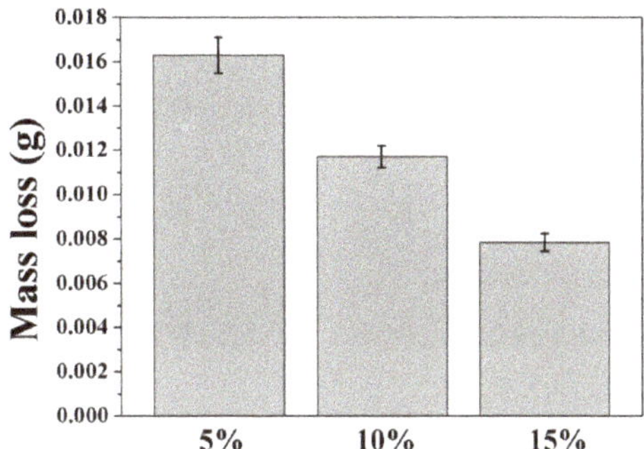

Figure 10. Mass loss of the NiCr–TiB_2–ZrB_2 coatings.

To better understand the wear behaviour of the NiCr–TiB_2–ZrB_2 composite coatings, SEM morphologies and EDS analyses (Table 4) of the worn surfaces are shown in Figure 11. As shown in Figure 11a, the worn surface of NiCr–5TiB_2–ZrB_2 was characterised by superficial and narrow grooves (zones A, D, and H), adhesive craters and a few cracks (zone B) with severe plastic deformation, oxygen element (zones C, E, G, and J) (from Table 4), TiB_2 phases (zones F and I) (from Table 4), and the white light zone. This implies that abrasive wear, adhesive wear, fatigue wear, and oxidation wear occurred.

During the test, the temperature between the coating and the counterbody increased gradually with a prolonged operation time, and thus, the adhesion between them may occur as a result of the direct contact. Through the research of Kliagua [27], this adhesion occurred between a ceramic material and cermet under a high temperature and high pressure. Therefore, the splats on the surface of the coating can adhere to the ball before being dragged out, which caused the emergence of the transfer layer. The peeling of the splats can be attributed to the limited interlamellar bonding, which was very common in the thermal-sprayed coatings [28]. The EDS analysis results showed that the white zones were oxidized with an oxygen content of up to 16.94 at.% (Table 4). The friction process released heat and caused an increase in the temperature on the upper sliding surface, which finally leads to oxidation. In addition, the oxidized phases can be crushed, and fine oxide particles were formed. Under the load applied, these fine oxide particles played the part of ploughs and the furrows along the sliding direction formed. It was reported that the fatigue wear resistance of the materials increased with the surface hardness [29]. Hence, when the hardness of the materials was lower, the probability of fatigue wear increased with the increasing wear time.

Many fragments of wear debris were also accumulated (Figure 11b–d), the morphologies of which were similar to the NiCr–5TiB_2–ZrB_2 coating, except for the cracks. However, the degree of adhesive wear gradually decreased, which may be because the hardness of the coatings rose significantly. In addition, the NiCr–15TiB_2–15ZrB_2 composite coating had the best tribological properties.

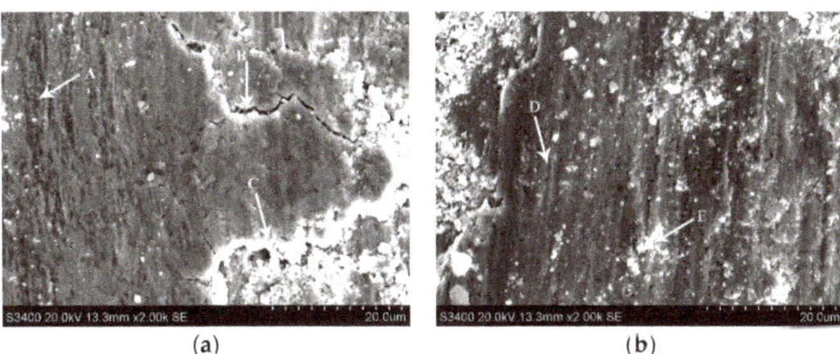

Figure 11. *Cont.*

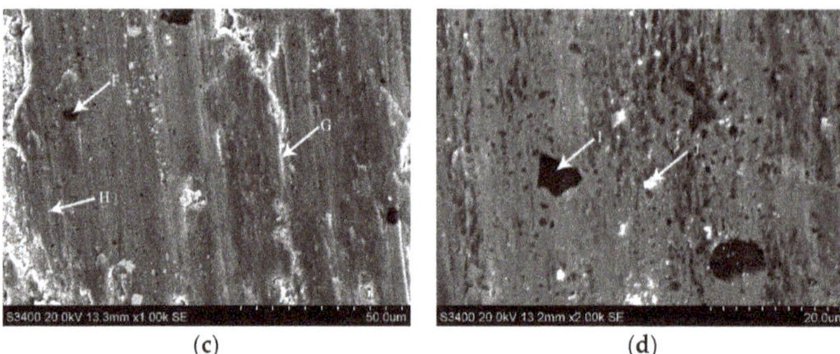

Figure 11. Worn morphologies of surface: (**a**) NiCr–5TiB_2–5ZrB_2; (**b**) NiCr–10TiB_2–10ZrB_2; (**c**) NiCr–15TiB_2–15ZrB_2; and (**d**) higher magnification image NiCr–15TiB_2–15ZrB_2.

Table 4. Chemical composition (at.%) in different regions in Figure 11.

Zone	Ni	Cr	Ti	Zr	B	O
C	47.86	12.69	2.86	1.29	18.36	16.94
E	42.89	10.43	4.07	3.11	25.87	13.63
F	–	–	33.54	–	66.46	–
I	–	–	34.73	–	65.27	–
J	35.97	8.54	5.66	3.29	34.72	11.82

4. Conclusions

The NiCr–TiB_2–ZrB_2 composite coatings were successfully fabricated on the SUS304 surface using high-energy ball milling and air plasma spraying technology. The microstructure and mechanical properties of the NiCr–TiB_2–ZrB_2 coatings with various TiB_2 and ZrB_2 contents were investigated. The following results can be highlighted:

- With an increase in the proportion of ceramics, the microhardness of the composite coating gradually increased. The average hardness obtained on the coating cross-section was 753, 957, and 1102 $HV_{0.1}$ for NiCr (5 wt.%)–TiB_2 (5 wt.%)–ZrB_2, NiCr (10 wt.%)–TiB_2 (10 wt.%)–ZrB_2, and NiCr (15 wt.%)–TiB_2 (15 wt.%)–ZrB_2 coatings.
- From the analyses of the Weibull distributions, the R value of all of the coatings was greater than 0.9 and there was little overall change of the coatings. With the addition of the ceramic phases (TiB_2 and ZrB_2), the β-values of the coatings rose from 67.39 to 134.28, implying that the coating had a minimal variation when the content of the ceramic phases (TiB_2 and ZrB_2) increased.
- The wear mechanism of the coatings was mixed, involving abrasive wear, adhesive wear, oxidation wear, and fatigue wear. The ceramic phases in the coatings gave outstanding resistance to wear and oxidation. The NiCr–15TiB_2–15ZrB_2 composite coating had the best tribological properties.

Author Contributions: Investigation, N.Z.; Conceptualization and Supervision, N.N.Z.; SEM (scanning electron microscope) Analysis, S.G.; XRD (X-ray diffraction) Analysis, S.L.; Data Reduction, G.Z.; Writing–Original Draft, N.Z; Writing–Review & Editing, Y.Z.

Funding: This research was funded by the Natural Science Foundation of Liaoning Province (201602553) and Chinese National Natural Science Foundation (51301112).

Acknowledgments: Thanks to Yingdong Qu for giving help on the wearing surface observation.

Conflicts of Interest: The authors declare no conflict of interest.

References

1. Alam, P.; Toivakka, M. Deflection and plasticity of soft-tip bevelled blades in paper coating operations. *Mater. Design* **2009**, *30*, 871–877. [CrossRef]

2. Hua, Y.; Xu, L.; Hua, W. Development of tungsten carbide blade. *Chin. Pulp. Pap.* **2013**, *32*, 49–52. [CrossRef]
3. Wang, H.; Li, H.; Zhu, H.; Cheng, F.; Wang, D.; Li, Z. A comparative study of plasma sprayed TiB_2-NiCr and Cr_3C_2-NiCr composite coatings. *Mater. Lett.* **2015**, *153*, 110–113. [CrossRef]
4. Zhang, C.; Luo, G.; Zhang, J.; Dai, Y.; Shen, Q.; Zhang, L. Synthesis and thermal conductivity improvement of W-Cu composites modified with WC interfacial layer. *Mater. Design* **2017**, *127*, 233–242. [CrossRef]
5. Zhang, Y.; Ma, J. Tungsten carbide ceramics. In *Practical Handbook of Ceramic Materials*, 1st ed.; Chemical Industry Press: Beijing, China, 2006; pp. 390–399. ISBN 7-5025-8469-2. (In Chinese)
6. Vashishtha, N.; Khatirkar, R.K.; Sapate, S.G. Tribological behaviour of HVOF sprayed WC-12Co, WC-10Co-4Cr and Cr_3C_2-25NiCr coatings. *Tribol. Int.* **2017**, *105*, 55–68. [CrossRef]
7. Zhu, H.B.; Li, H.; Li, Z.X. Plasma sprayed TiB_2-Ni cermet coatings: Effect of feedstock characteristics on the microstructure and tribological performance. *Surf. Coat. Technol.* **2013**, *235*, 620–627. [CrossRef]
8. Zhang, Y.; Yu, F.; Hao, S.; Dong, F.; Xu, Y.; Geng, W.; Zhang, N.; Gey, N.; Grosdidier, T.; Dong, C. Evolution of nanostructure and metastable phases at the surface of a HCPEB-treated WC-6% Co hard alloy with increasing irradiation pulse numbers. *Coatings* **2017**, *7*, 178. [CrossRef]
9. Zhang, M.; Qu, K.; Luo, S.; Liu, S. Effect of Cr on the microstructure and properties of TiC-TiB_2 particles reinforced Fe-based composite coatings. *Surf. Coat. Technol.* **2017**, *316*, 131–137. [CrossRef]
10. Yan, H.; Zhang, P.; Yu, Z.; Lu, Q.; Yang, S.; Li, C. Microstructure and tribological properties of laser-clad Ni-Cr/TiB_2, composite coatings on copper with the addition of CaF_2. *Surf. Coat. Technol.* **2012**, *206*, 4046–4053. [CrossRef]
11. Lotfi, B.; Shipway, P.H.; Mccartney, D.G.; Edris, H. Abrasive wear behaviour of Ni(Cr)-TiB_2 coatings deposited by HVOF spraying of SHS-derived cermet powders. *Wear* **2003**, *254*, 340–349. [CrossRef]
12. Jones, M.; Horlock, A.J.; Shipway, P.H.; McCartney, D.G.; Wood, J.V. A comparison of the abrasive wear behavior of HVOF sprayed titanium carbide-and titanium boride-based cermet coatings. *Wear* **2001**, *251*, 1009–1016. [CrossRef]
13. Xu, J.; Zou, B.; Zhao, S.; Hui, Y.; Huang, W.; Zhou, X.; Wang, Y.; Cai, X.; Cao, X. Fabrication and properties of ZrC-ZrB_2/Ni cermet coatings on a magnesium alloy by atmospheric plasma spraying of SHS powders. *Ceram. Int.* **2014**, *40*, 1537–15544. [CrossRef]
14. Umanskyi, O.; Poliarus, O.; Ukrainets, M.; Antonov, M.; Hussainova, I. High temperature sliding wear of NiAl-based coatings reinforced by borides. *Mater. Sci.* **2016**, *22*, 49–53. [CrossRef]
15. Farhadinia, F.; Sedghi, A. Wear behaviour of Al/(Al_2O_3 + ZrB_2 + TiB_2) hybrid composites fabricated by hot pressing. *Int. J. Mater. Res.* **2015**, *106*, 160–165. [CrossRef]
16. Chakraborty, S.; Debnath, D.; Mallick, A.R.; Das, P.K. Mechanical and thermal properties of hot pressed ZrB_2, system with TiB_2. *Int. J. Refract. Met. Hard Mater.* **2014**, *46*, 35–42. [CrossRef]
17. Li, B. Effect of ZrB_2 and SiC addition on TiB_2-based ceramic composites prepared by spark plasma sintering. *Int. J. Refract. Met. Hard Mater.* **2014**, *46*, 84–89. [CrossRef]
18. Luo, P.; Xiong, C.; Wang, C.; Qin, F.; Xiao, Y.; Li, Z.; Liu, K.; Guan, X.; Dong, S. Comparative failure analysis of electrodes coated with TiB_2-ZrB_2 and TiB_2-ZrB_2/Ni layers. *Surf. Coat. Technol.* **2017**, *317*, 83–94. [CrossRef]
19. Zhang, L.; Wang, C.; Qian, S.; Yu, Q.; Dong, C. Microstructure and wear resistance of laser-clad $(Co,Ni)_{61.2}B_{26.2}Si_{7.8}Ta_{4.8}$ coatings. *Metals* **2017**, *7*, 419. [CrossRef]
20. Wu, Y.; Zeng, D.; Liu, Z.; Qiu, W.; Zhong, X.; Yu, H.; Li, S. Microstructure and sliding wear behavior of nanostructured Ni60–TiB_2 composite coating sprayed by HVOF technique. *Surf. Coat. Technol.* **2011**, *206*, 1102–1108. [CrossRef]
21. Zhang, N.; Lin, D.; He, B.; Zhang, G.; Li, D. Effect of oxyacetylene flame remelting on wear behaviour of supersonic air-plasma sprayed NiCrBSi/h-BN composite coatings. *Surf. Rev. Lett.* **2017**, *24*, 83–90. [CrossRef]
22. Pawlowski, L. Coating growth. In *The Science and Engineering of Thermal Spray Coatings*, 1st ed.; Li, H., He, D., Eds.; China Machine Press: Beijing, China, 2011; pp. 135–136. ISBN 978-7-111-32163-7. (In Chinese)
23. Zhang, N.; Zhang, N.; Wei, X.; Zhang, Y.; Li, D. Microstructure and tribological performance of TiB_2-NiCr composite coating deposited by APS. *Coatings* **2017**, *7*, 238. [CrossRef]
24. Zhu, H.; Li, H.; Yang, H.; Li, Z. Microstructure and sliding wear performance of plasma-sprayed TiB_2-Ni coating deposited from agglomerated and sintered powder. *J. Therm. Spray. Technol.* **2013**, *22*, 1310–1319. [CrossRef]

25. An, Y.; Li, S.; Hou, G.; Zhao, X.; Zhou, H.; Chen, J. Mechanical and tribological properties of nano/micro composite alumina coatings fabricated by atmospheric plasma spraying. *Ceram. Int.* **2017**, *43*, 5319–5328. [CrossRef]
26. Zhang, F.; He, J.; Chen, K.; Qin, Y.; Li, C.; Yin, F. Microstructure evolution and mechanical properties of TiCN-Cr nano/micro composite coatings prepared by reactive plasma spraying. *Appl. Surf. Sci.* **2017**, *427*, 905–914. [CrossRef]
27. Kliagua, A.M.; Travessa, D.; Ferrante, M. Al_2O_3/Ti interlayer/AISI 304 diffusion bonded joint: Microstructural characterization of the two interfaces. *Mater. Charact.* **2001**, *46*, 65–74. [CrossRef]
28. Tian, L.; Feng, Z.; Xiong, W. Microstructure, microhardness, and wear resistance of AlCoCrFeNiTi/Ni60 coating by plasma spraying. *Coatings* **2018**, *8*, 112. [CrossRef]
29. Lu, Z.; Zhou, Y.; Rao, Q.; Jin, Z. An investigation of the abrasive wear behavior of ductile cast iron. *J. Mater. Process. Technol.* **2001**, *116*, 176–181. [CrossRef]

Article

Microstructure, Microhardness, and Wear Resistance of AlCoCrFeNiTi/Ni60 Coating by Plasma Spraying

Lihui Tian *, Zongkang Feng and Wei Xiong

National Demonstration Center for Experimental Materials Science and Engineering Education, Jiangsu University of Science and Technology, Zhenjiang 212003, China; zongkang_feng@163.com (Z.F.); xiongwei32410@163.com (W.X.)

* Correspondence: tianlihui@just.edu.cn; Tel.: +86-511-8440-1184

Received: 7 February 2018; Accepted: 15 March 2018; Published: 19 March 2018

Abstract: In this comparative study, Ni60 was used as a reinforcement and added into an AlCoCrFeNiTi high-entropy alloy (HEA) matrix coating in order to enhance its hardness and wear resistance. An AlCoCrFeNiTi/Ni60 coating was prepared by plasma spraying with mechanically-blended AlCoCrFeNiTi/Ni60 powder. The coating microstructure was observed and analyzed. Bonding strength, microhardness, and wear resistance of the coating were investigated. The results showed that a compact AlCoCrFeNiTi/Ni60 coating with Ni60 splats uniformly distributed in the AlCoCrFeNiTi matrix was deposited. After spraying, matrix body-centered cubic (BCC), face-centered cubic, and ordered BCC phases were detected in the coating. The added large Ni60 particles played an important role in strengthening the coating. Tensile test results showed that bonding strength of this coating was above 60.1 MPa, which is far higher than that of the AlCoCrFeNiTi HEA coating in the previous study. An average microhardness of 676 HV was obtained for the main body of the coating, which is much higher than that of the AlCoCrFeNiTi HEA coating. Solution hardening in γ-Ni and dispersion strengthening of the hard interstitial compounds, such as Cr_7C_3, CrB, Cr_2B, and $Cr_{23}C_6$ increased the hardness of Ni60, and then the AlCoCrFeNiTi/Ni60 coating. During wear testing at 25 °C, adhesive wear and abrasive wear occurred, while, at 500 °C, abrasive wear took place. Volume wear rates of the coating at 25 °C and 500 °C were $0.55 \pm 0.06 \times 10^{-4}$ $mm^3 \cdot N^{-1} \cdot m^{-1}$ and $0.66 \pm 0.02 \times 10^{-4}$ $mm^3 \cdot N^{-1} \cdot m^{-1}$, respectively. Wear resistance of this coating was better than that of the AlCoCrFeNiTi HEA coating, which can be attributed to addition of the hard Ni60. Therefore, Ni60 is an appropriate reinforcement to further enhance the wear resistance of HEA coating.

Keywords: high-entropy alloy (HEA) matrix coating; plasma spraying; wear resistance; microhardness; bonding strength

---

1. Introduction

High-entropy alloys (HEAs), which are usually composed of at least five principal metallic elements in equimolar or near-equimolar ratios, were defined by Yeh et al. in 2004 [1]. Due to their outstanding properties (e.g., high strength, sufficient hardness, and excellent wear resistance) [2–6], HEAs are a kind of promising coating materials used in harsh environments, especially at high temperature. So far, most of the HEA coatings have been prepared by electrochemical deposition [7], magnetron sputtering [8,9], and laser cladding [10,11]. However, the preparation and industrial application of HEA coatings are limited owing to the low deposition efficiency, high residual stress, high dilution, and high cost of these techniques.

Compared with the above coating techniques, plasma spraying, which has been widely used to prepare coatings because of its low production cost and high efficiency, is an appropriate method to deposit HEA coatings. However, few investigations on plasma-sprayed HEA coatings have been done until now [12,13]. An AlCoCrFeNi coating with a Vickers hardness of about 421 HV and a MnCoCrFeNi coating with that of about 451 HV were deposited by plasma spraying [12]. Plasma-sprayed $Ni_xCo_{0.6}Fe_{0.2}Cr_ySi_zAlTi_{0.2}$ HEA coatings were investigated and a microhardness of 450 ± 30 HV was obtained for the $NiCo_{0.6}Fe_{0.2}Cr_{1.5}SiAlTi_{0.2}$ coating [13].

In addition, in the authors' previous study [14], an AlCoCrFeNiTi HEA coating (hereinafter referred to as HEA coating) was deposited by plasma spraying using ball-milled powder as a feedstock. Investigation results showed that a higher bonding strength of 50.3 MPa and an average microhardness of 642 HV were obtained for the coating at 25 °C. Meanwhile, the coating also exhibited better wear resistance at higher temperatures of 700 °C and 900 °C. However, at test temperatures of 25 °C and 500 °C, the volume wear rates were $0.77 \pm 0.01 \times 10^{-4}$ $mm^3 \cdot N^{-1} \cdot m^{-1}$ and $0.93 \pm 0.02 \times 10^{-4}$ $mm^3 \cdot N^{-1} \cdot m^{-1}$, respectively. These results indicate that the wear resistance of the coating at lower temperatures was not satisfactory.

For the plasma-sprayed AlCoCrFeNiTi protective coating on components of industrial machineries, improvement of its performances, such as hardness and wear resistance is quite necessary to meet the strict requirements of much severer conditions. Generally, to design and prepare a composite, i.e., adding a secondary material into the matrix is an effective approach to further enhance the hardness and wear resistance of the matrix materials. Until now, investigations on high-entropy alloy matrix composite materials and coatings have not been reported.

Usually, oxides, nitrides, borides, and carbides are chosen as reinforcements of composites [15,16]. However, owing to their brittleness, the toughness of the composites can be deteriorated significantly. As a conventional coating material with high hardness, sufficient strength, excellent wear, and heat resistances, Ni-based self-fluxing alloy has been widely used in fields of petroleum and chemical industries, power, and national defense [17,18]. By now, a large number of coating technologies, such as overlay welding, laser cladding, flame spraying, high velocity oxy-fuel spraying, and plasma spraying have been employed to deposit Ni-based self-fluxing alloy coatings [17–19]. In addition to the high hardness and excellent wear resistance at both room temperature and high temperature, good toughness makes Ni-based self-fluxing alloy a more appropriate reinforcement than the brittle traditional ones. However, no studies on composite materials and coatings reinforced by Ni-based self-fluxing alloy have yet been reported.

Therefore, in this comparative study, to enhance the hardness and wear resistance of the AlCoCrFeNiTi coating, Ni60 was used as a reinforcement and added into the matrix. An AlCoCrFeNiTi/Ni60 coating (hereinafter referred to as HEA/Ni60 coating) was prepared by plasma spraying. Mechanically-blended AlCoCrFeNiTi/Ni60 powder (hereinafter referred to as HEA/Ni60 powder) was used as a feedstock. The microstructure of the plasma-sprayed coating was observed and analyzed. Bonding strength, microhardness, and wear resistance at 25 °C and 500 °C of the coating were comparatively investigated.

2. Experimental Materials and Procedures

2.1. Materials

In the present study, AlCoCrFeNiTi powder (hereinafter referred to as HEA powder) was prepared by ball milling using equimolar related elemental powders with a size of less than 75 μm as starting materials. Ball milling was conducted with a planetary ball mill (XM-4, Sanxing Instruments Ltd., Xiangtan, China). Both the pots and the balls were made of 304 stainless steel. More details about the ball milling process can be found in our previous study [14]. The powder milled for 30 h was mechanically blended with a commercial atomized Ni60 powder (30–80 μm) with an AlCoCrFeNiTi/Ni60 weight ratio of 7:3. Chemical composition of the Ni60 powder is listed in Table 1.

Table 1. Chemical composition of Ni60 powder (wt %).

Element	Content
C	0.63
B	2.97
Si	4.50
Fe	5.31
Cr	16.20
Ni	Bal.

2.2. Coating Preparation

316 stainless steel discs with a diameter of 25 mm and a thickness of 8 mm were used as substrates. To increase the coating bonding strength, the substrates were cleared with alcohol and sandblasted with alumina (<1 mm) before coating deposition. Sandblasting was carried out with the following parameters: compressed air pressure (0.50 MPa), blasting time (30 s), blasting angle (90°). A plasma spraying system (PRAXAIR-3710, Praxair Surface Technologies, Indianapolis, IN, USA) was employed to prepare the HEA/Ni60 coating. Plasma spraying was done with the following parameters: plasma arc power (45 kW), main gas (Ar) pressure (0.41 MPa), secondary gas (N_2) pressure (0.45 MPa), powder carrier gas (Ar) pressure (0.28 MPa), rotating speed of powder feeder (0.8 $r \cdot min^{-1}$), spray distance (100 mm), and gun traverse speed (200 $mm \cdot s^{-1}$).

2.3. Characterization

An X-ray diffractometer (XRD, LabX XRD-6000, SHIMADZU, Kyoto, Japan) equipped with a Cu target operating at 1.2 kW was employed to measure the phases present in the feedstock and the coating. During the measurement, the scanning step was 0.02° and the scanning speed was $6° \cdot min^{-1}$.

To observe and analyze surface morphology of the feedstock, cross-sectional microstructure of the coating, and morphology of the worn surfaces, a field emission scanning electron microscopy (FESEM, ZEISS ΣIGMA, ZEISS, Oberkochen, Germany) equipped with an energy dispersive spectroscopy (EDS, x-act, OXFORD INSTRUMENTS, Oxford, UK) was used. The accelerating voltage of the scanning electron microscope was 20 kV.

Tensile experiments according to the standard ISO 14916:1999 [20] were carried out to measure the coating bonding strength, and the sketch of the measurement can be found in the previous paper [14]. E-7 adhesive was used to cement the coating specimens and the blocks. A tensile speed of 1 $mm \cdot min^{-1}$ was chosen. Three specimens were measured and the average value of the tensile strength was used to evaluate the bonding strength of the coating.

An MH-5 mode Vickers microhardness tester (Shanghai Everone Precision Instruments Ltd., Shanghai, China) was used to measure the coating hardness along the direction perpendicular to the coating/substrate interface. During the measurement, a load of 200 g and dwelling time of 10 s were used. For each distance from the interface, more than 10 points were measured and the average values were calculated.

A high-temperature ball-on-disc friction and wear tester (HT-1000, Zhongke Kaihua Science and Technology Development Ltd., Lanzhou, China) was used to investigate the wear resistance of the coating. Figure 1 shows the schematic diagram of the wear test. Si_3N_4 balls with a diameter of Φ 5 mm were fixed on a loading rod. The coating specimen was fixed on a specimen pan by screws. They can contact with each other with the help of the weight with a load of 5 N. During the test, friction force was continuously measured by a force sensor, and then divided by the normal load to calculate the friction coefficient.

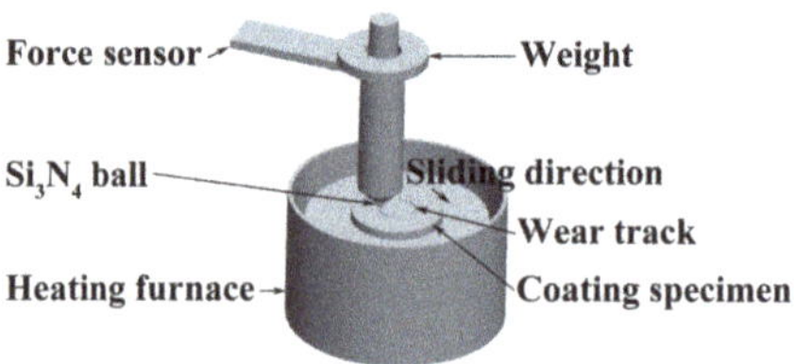

Figure 1. Schematic diagram of wear test.

Before testing, the coating was polished to reach a roughness of about R_a = 0.7 μm. When the load was added, the heating furnace was heated up to test temperatures with a heating speed of 10 °C· min^{-1} in atmospheric environment. The coating specimen rotated at a set velocity of 573 r·min^{-1} with a friction radius of 5 mm. The coating specimen was taken out when the wear time reached 20 min. A three-dimensional (3D) confocal laser scanning microscope (LEXT OLS4000, OLYMPUS, Tokyo, Japan) was used to measure the cross-sectional area of the wear tracks A (mm^2). The volume wear rate W ($mm^3 \cdot N^{-1} \cdot m^{-1}$) was calculated according to $W = A \cdot P/(S \cdot L)$, in which P is perimeter of the wear track in mm, S is sliding distance in m, and L is the load in N. For each wear temperature, three specimens were tested and the volume wear rate result was reported as an average value. More details about the wear test were described in previous publications [14,21].

3. Results and Discussion

3.1. Microstructure of Feedstock and HEA/Ni60 Coating

Figure 2a shows the surface morphology of the mechanically blended HEA/Ni60 powder. It can be observed that the atomized Ni60 particles were spherical in shape with a smooth surface, while the ball-milled AlCoCrFeNiTi particles were near-equiaxed in shape with a rough surface. At a higher magnification (Figure 2b), gaps can be observed on the surface of the AlCoCrFeNiTi particle, and small particles adhered to it. These characteristics were caused by the particles which were not well cold-welded during the ball milling process. Due to its appropriate particle size and shape, the mixed powder exhibited good flowability and was suitable for plasma spraying.

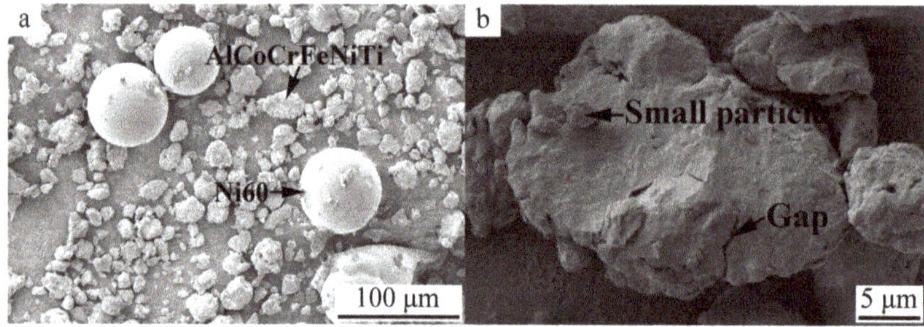

Figure 2. Surface morphology of feedstock: (**a**) mechanically blended HEA/Ni60 powder and (**b**) HEA powder at a higher magnification.

Figure 3 provides the XRD patterns of the mechanically blended HEA/Ni60 powder and the coating. From the pattern of the feedstock, phases such as γ-Ni, Cr_7C_3, CrB, Cr_2B, Ni_3B, and $Cr_{23}C_6$ present in the atomized Ni60 powder were detected. Additionally, diffraction peaks of typical body-centered cubic (BCC) phase at 44.6°, 65.0°, and 82.3° were detected, which were obviously broadened because of the grain refinement and increase of the lattice strain of the ball milled HEA powder during ball milling.

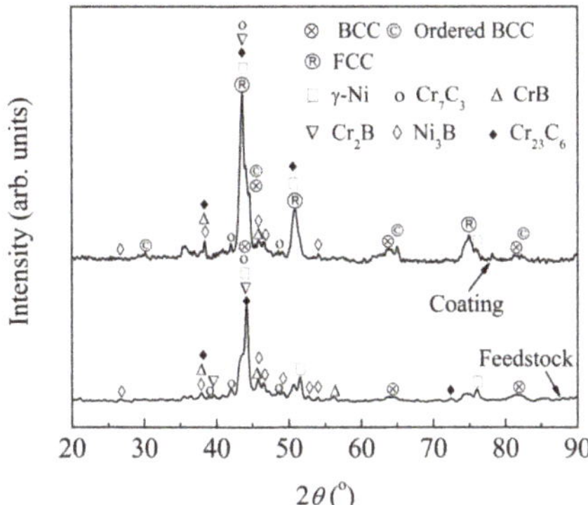

Figure 3. X-ray diffraction (XRD) patterns of feedstock and HEA/Ni60 coating.

As shown in Figure 3, in the plasma-sprayed coating, the peak intensity of the matrix BCC phase decreased, and new phases, such as face-centered cubic (FCC) and minor ordered BCC phases were found. This can resulted from the formation of the stable equilibrium phases from the metastable supersaturated solid solution present in the ball milled HEA feedstock [4]. Moreover, compared with that of the feedstock, a slight shift to the left of the major diffraction peak can be observed in Figure 3, which was due to the difference of crystallinity level and lattice strain between the feedstock and the coating [22]. Furthermore, no obvious phase transformation of Ni60 occurred after plasma spraying.

Figure 4 provides cross-sectional microstructure of the HEA/Ni60 coating. From Figure 4a, a typical lamellar structure of a plasma-sprayed coating containing major molten flattened splats and minor unmelted particles can be observed. Ni60 splats with a thickness of less than 40 μm were distributed uniformly in the AlCoCrFeNiTi matrix. Very few pores present in the coating with a size of less than 20 μm. It was reported that these pores are caused by imperfect wetting and inadequate filling of the melting droplets to the previously formed rough splat surfaces [23–26]. Apart from these pores, the coating was compact.

EDS analysis was conducted on points with typical contrasts on the coating cross-section in Figure 4b. The chemical composition is listed in Table 2. Point A, which was rich in Co, Cr, Fe, and Ni, was related to the BCC and ordered BCC phases (Figure 3) presented in the AlCoCrFeNiTi high-entropy alloy splats. Point B was rich in Co and Ni, but not Al, and it was related to the FCC phase in the HEA splat detected by XRD. According to the previous reports [27,28], Cr and Ni are promoters of BCC and FCC phases, respectively. It was also reported that BCC and ordered BCC phases which were rich in Fe-Cr and Al-Ni were found by EDS in vacuum induction melting $AlCoCrFeNiTi_{0.5}$ HEAs [29]. In the Ni60 splats, the matrix (e.g., point C) with a light gray contrast and a large amount of Ni and Cr was identified to be the γ-Ni solid solution phase detected by XRD (Figure 3). The cellular eutectic precipitations with a dark gray contrast (e.g., point D) scattered in the matrix could be the reinforcements (such as CrB, Cr_2B, and Ni_3B phases) due to its high content of B. Additionally, a few points (e.g., point E) with a dark gray contrast and higher oxygen content were caused by the oxidation of HEA powder owing to the air permeation into the plasma jet [30]. Usually, during plasma spraying, because of the operating atmosphere, oxidation of the metallic elements cannot be avoided [31]. However, diffraction peaks of these oxides were not detected in Figure 3 due to their lower contents.

Figure 4c shows the coating/substrate interface at a higher magnification. It is noted that the surface profile of the substrate in the HEA/Ni60 coating is much rougher than that in the HEA coating [14]. Thus, during spraying, as the high-temperature droplets impact on the substrate surface, and then spread along the rough surface, strong mechanical bonding can be formed due to the cooling, solidification and shrinkage of the splats. In addition, the addition of large Ni60 particles played an important role in strengthening the coating. First, Ni60 with a lower melting point can completely wet along the previously formed rough surface and sufficiently fill in the holes, which resulted in a strong bonding between the Ni60 particles and the nether splats. Second, since Ni60 is present in the coating as large particles, they can increase the surface roughness of the formed coating after their deposition. Therefore, strong mechanical bonding can be formed between the subsequent splats and the previously formed Ni60 particles. Thus, good contact between Ni60 particles and AlCoCrFeNiTi splats was formed (as shown in Figure 4c), which can cause a higher cohesion strength of the coating. Indeed, tensile test results showed that fracture of the specimen mainly occurred in the E-7 adhesive layer (Figure 5) and the average tensile strength was 60.1 MPa. Therefore, bonding strength of this coating was above 60.1 MPa, which is far higher than that of the HEA coating (50.3 MPa) in the previous study [14].

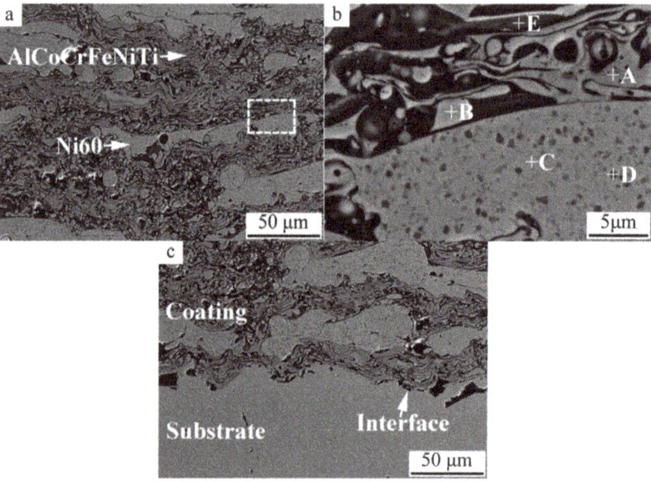

Figure 4. Cross-sectional microstructure of HEA/Ni60 coating (**a**,**b**) and coating/substrate interface (**c**).

Table 2. Energy dispersive spectroscopy (EDS) analysis results of typical points in Figure 4b (at %).

Point	Al	Co	Cr	Fe	Ni	Ti	Si	O	C	B
A	8.66	21.65	17.72	20.16	20.98	10.83	0.00	0.00	0.00	0.00
B	0.00	33.09	1.29	5.12	53.95	0.94	0.00	5.61	0.00	0.00
C	0.00	0.00	39.75	4.94	49.47	0.00	5.32	0.00	0.52	0.00
D	0.00	0.00	29.21	8.44	25.76	0.00	4.49	0.00	0.00	32.10
E	6.28	7.24	1.94	18.42	7.77	4.42	0.00	53.93	0.00	0.00

Figure 5. Fracture of tensile test specimen.

3.2. Microhardness of HEA/Ni60 Coating

Microhardness of the HEA/Ni60 coating is shown in Figure 6, and that of the HEA coating in our previous study [14] are also given out for comparison. As shown in Figure 6, microhardness of the HEA/Ni60 coating increased significantly from the substrate to the substrate/coating interface, and then to the coating. For example, at the distance of −100 μm, the substrate microhardness was 183 HV, while at the distance of −50 μm, it increased to 289 HV. A value of 443 HV was obtained at the substrate/coating interface. The coating microhardness was 636 HV, 665 HV, and 727 HV at the distance of 50 μm, 100 μm, and 150 μm, respectively. At the distance of 200 μm (near the coating surface), a lower value of 639 HV was obtained.

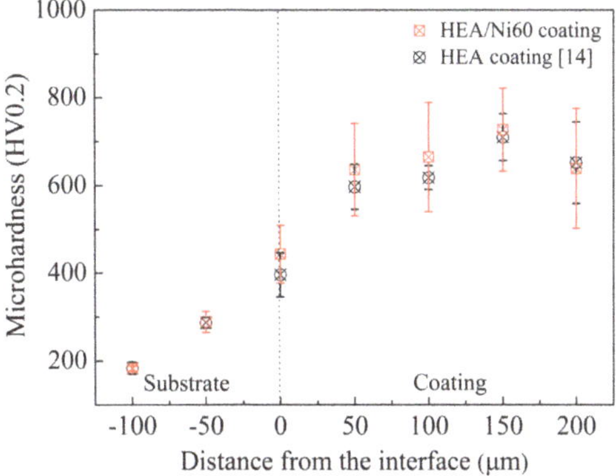

Figure 6. Microhardness of HEA/Ni60 coating.

For both the HEA/Ni60 and HEA coatings, a similar microhardness was obtained at distances of −100 μm and −50 μm. This can be attributed to the same sand blasting parameters. During sand blasting, work hardening took place [31], and this can increase the substrate hardness significantly to a value of nearly 300 HV, e.g., at the distance of −50 μm. However, at a deeper distance (e.g., −100 μm), the substrate hardness was not affected significantly, and it was close to that of 316 stainless steel prior to spraying (173 HV). At the substrate/coating interface, the hardness in this study was much higher than that of the HEA coating (396 HV). It can be attributed to the better mechanical bonding in Figure 4c, which also caused a higher bonding strength.

An average microhardness of 676 HV was obtained for the main body of the coating, which is far higher than that of the HEA coating in the previous work (642 HV) [14]. The main reason can be the addition of the reinforcement, Ni60. As discussed in Section 3.1, after spraying, γ-Ni, Cr_7C_3, CrB, Cr_2B, and $Cr_{23}C_6$ phases were detected in the coating. For the γ-Ni solid solution, solution hardening can strength the Ni60 splat significantly. Moreover, it was reported that for Ni60, the addition of Cr element can improve its hardness by formation of carbides and borides with high hardness, and the addition of B element can enhance the formation of the hard phases [19]. Thus, the dispersion strengthening of the hard interstitial compounds such as Cr_7C_3, CrB, Cr_2B, and $Cr_{23}C_6$ can also strength the Ni60 splat. Both the solution hardening and dispersion strengthening increased the hardness of Ni60, and then the HEA/Ni60 coating. Furthermore, porosity has significant effect on the hardness of thermally-sprayed coatings [31]. As shown in Figure 4a, the addition of the large Ni60 particles with a low melting point can completely wet along the previously surface and fill in the holes, and finally result in a good bonding between the molten Ni60 and AlCoCrFeNiTi splats, and a compact coating was obtained. Hence, the low porosity of the coating in the present study can be another important reason of the high hardness. At the distance of 200 μm, the coating hardness decreased to a lower value. This can be resulted from the edge effect of thermally-sprayed coatings. During spraying, lower effect of sintering induced by few subsequent particles can increase the porosity of the coating [32], which can decrease the coating hardness.

3.3. Wear Resistance of HEA/Ni60 Coating

3.3.1. Wear Mechanism

To investigate the wear mechanism of the HEA/Ni60 coating, morphology of the worn surfaces at 25 °C and 500 °C was analyzed.

Figure 7a,c show the morphology of worn surface of the coating and the counterpart at 25 °C. From Figure 7a, an average width of about 0.76 mm for the wear track was obtained. The worn surface of the coating was quite rough with large numbers of delamination and lips present on it (Figure 7a,b), and lots of fine flakes with a dimension of less than 10 μm along the sliding direction can be observed on the counterpart surface (Figure 7c). These facts illustrate that adhesive wear took place during the wear experiment. During testing, welding between the coating and the counterpart occurred due to their direct contact under the load applied. As the ball slid on the coating, splats on the surface of the coating can adhere to the ball, and then be dragged out, which results in delamination. Peeling of the splats was caused by the limited interlamellar bonding ratio. In thermally-sprayed coatings, the bonding ratio between flattened splats was investigated, and a result of less than one third was obtained [31]. Consequently, the peelings can be pushed to other areas of the worn surface along the sliding direction, and the lips formed. Repeated adhesion and shear deformation accelerated the formation of lips and flakes. EDS analysis result showed that the lips (e.g., point 1 in Figure 7b) were oxidized with an oxygen content of up to 64.57 at % (Table 3). It was reported that due to the friction heating, flash temperature on the up-most sliding surface (e.g., lips) would increase [33], which caused the oxidation of the lips. The solid solution matrix of BCC, ordered BCC, FCC, and γ-Ni with excellent plasticity and toughness prevented the moving of the lips, which can further enhance the coating wear resistance. In addition, lots of furrows with a width of about 10 μm can be observed on the worn surface in Figure 7b, which illustrates that abrasive wear was another wear mechanism. During wear testing, the hard oxidized lips can be crushed, and fine oxide particles formed. Under the load applied, these fine oxide particles play the part of ploughs and the furrows along the sliding direction were formed. It is worth noting that on the surface of Ni60 (e.g., point 2 in Figure 7b), which was identified by EDS (Table 3), no obvious furrows can be seen. This can be explained by the high hardness of Ni60 at 25 °C, which is much higher than that of AlCoCrFeNiTi [14]. The characteristics of delamination, lips, and furrows cause a higher roughness of the worn surface.

Compared with that at 25 °C, the morphology of worn surface of the coating and the counterpart changed significantly at 500 °C (Figure 7d–f). With increasing the wear temperature from 25 to 500 °C, the wear track became narrower (Figure 7d), with an average width of about 0.59 mm. A smoother wear surface was obtained and no obvious delamination and lips are observed. At the edge of the wear track (Figure 7d), obvious plastic deformation can be observed. Additionally, the flakes adhering to the counterpart became a thin slice with a dimension from several micrometers to tens of micrometers (Figure 7f). As the temperature increased to 500 °C, the coating hardness decreased, while the plasticity increased. During testing, as the ball slid on the coating surface, plastic deformation of the metallic splats occurred with the load applied. Thus, the interlamellar gaps can be healed and good bonding between the splats formed, and delamination cannot occur and lips cannot form. Thus, a smoother wear surface was obtained. Due to the low hardness and good plasticity, plastic deformation occurred on the edge of the wear track. Moreover, finer furrows with a width of about 5 μm can be found in Figure 7e, which illustrates that abrasive wear took place. According to the EDS analysis results listed in Table 3, the coating surface (e.g., point 4) was slightly oxidized. Oxidation of the coating was attributed to the flash temperature during wear test and the defects existed in the coating (e.g., interlamellar gaps and pores), which is a typical problem for the plasma-sprayed metallic coatings. During high temperature wear, crushing of the tribo film on the worn surface took place, which resulted in the formation of fine particles with small sizes. Under the load applied, these fine particles play the part of ploughs and the furrows along the sliding direction were formed (Figure 7d,e). From Table 3, it can be found that point 3 was related to Ni60, which was not oxidized due to its oxidation resistance at 500 °C. However, furrows can also be observed on the surface of Ni60 particles. This could be owing to the decrease of its hardness. The coating material from the furrows was aggregated, extruded, and adhered to the surface of the counterpart as flakes with thin slice shapes because of its good plasticity at 500 °C.

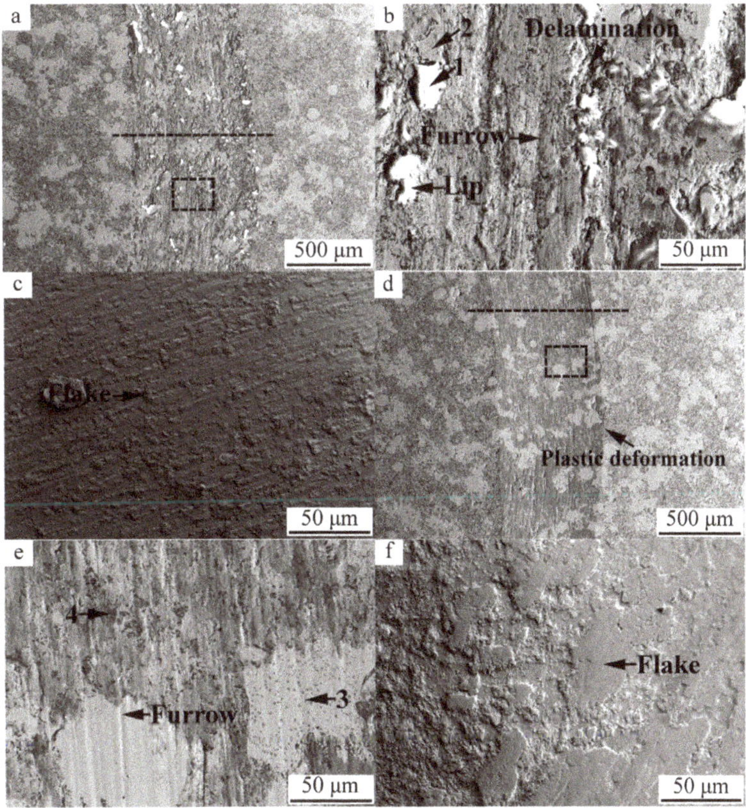

Figure 7. Worn surfaces of HEA/Ni60 coating and counterpart at 25 °C (**a–c**) and 500 °C (**d–f**).

Table 3. Energy dispersive spectroscopy analysis results of typical points in Figure 7 (at %).

Point	Al	Co	Cr	Fe	Ni	Ti	Si	O
1	2.39	3.95	4.05	3.92	11.58	5.04	4.50	64.57
2	0.00	0.00	25.15	5.10	47.60	0.50	5.67	15.98
3	0.00	0.00	15.46	4.93	71.57	0.00	8.04	0.00
4	3.07	6.08	4.59	6.03	12.52	6.91	0.87	59.92

3.3.2. Friction Coefficient

Figure 8 provides the friction coefficient of HEA/Ni60 coating as a function of wear time. From the wear curves, two obvious stages (running-in and steady wear stages) can be observed. For the curve at 25 °C, the friction coefficient increased within 5 min at the running-in stage. In the steady wear stage, the average friction coefficient is much higher with a larger fluctuation (0.80 ± 0.03), which can be attributed to the changing of the roughness from a polished smooth surface (with a roughness of about R_a = 0.7 μm) to a rough worn surface (Figure 7a,b). At high temperature, the friction coefficient curves showed short time running-in stages, which can be caused by the welding of two contact surfaces at high temperatures. At 500 °C, the average friction coefficient in the steady wear stage became lower with small fluctuation (0.57 ± 0.03), which was caused by the smooth, worn surfaces in Figure 7d,e.

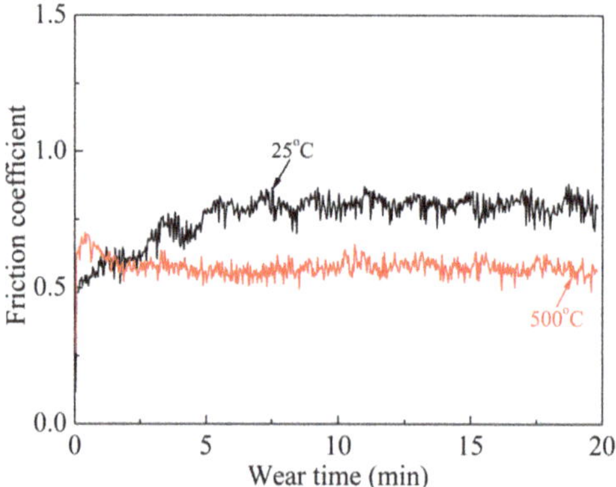

Figure 8. Friction coefficients of HEA/Ni60 coating at 25 °C and 500 °C.

3.3.3. Volume Wear Rate

Three-dimensional non-contact surface mapping across the wear tracks of HEA/Ni60 coating at 25 °C and 500 °C is shown in Figure 9. Figure 10 provides the volume wear rates of the HEA/Ni60 and HEA coatings as a function of test temperature. With increasing the wear temperature from 25 to 500 °C, the volume wear rates of the coating increased. For example, volume wear rates of the coating at 25 °C and 500 °C were 0.55 ± 0.06 × 10^{-4} $mm^3 \cdot N^{-1} \cdot m^{-1}$ and 0.66 ± 0.02 × 10^{-4} $mm^3 \cdot N^{-1} \cdot m^{-1}$, respectively. This could be owing to the decrease of the coating hardness. Wear resistance of this coating was better than that of the HEA coating and also the 316 stainless steel substrate material, which can result from the addition of the hard Ni60. Therefore, Ni60 is an appropriate reinforcement to further enhance the wear resistance of HEA coating.

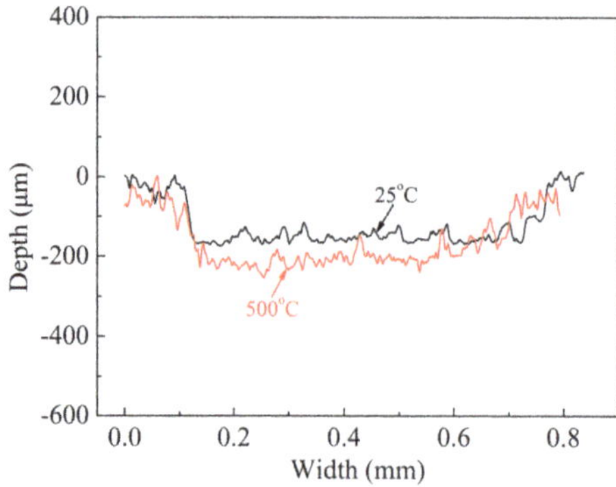

Figure 9. Three-dimensional (3D) non-contact surface mapping across the wear tracks of HEA/Ni60 coating at 25 °C and 500 °C in Figure 7a,d.

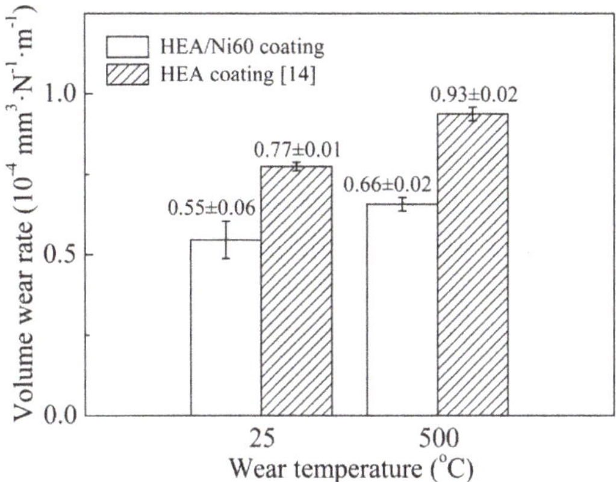

Figure 10. Volume wear rates of HEA/Ni60 coating at 25 °C and 500 °C.

4. Conclusions

A compact AlCoCrFeNiTi/Ni60 coating with Ni60 splats uniformly distributed in AlCoCrFeNiTi matrix was deposited by plasma spraying. After spraying, matrix BCC, FCC, and ordered BCC phases were detected in the coating.

The added large Ni60 particles played an important role in strengthening the coating. A strong mechanical bonding between Ni60 and AlCoCrFeNiTi splats caused a higher cohesion strength. Tensile test results showed that bonding strength of this coating was above 60.1 MPa, which is far higher than that of the AlCoCrFeNiTi coating in the previous study.

An average microhardness of 676 HV was obtained for the main body of the coating, which is far higher than that of the HEA coating in the previous work. The main reason can be attributed to the addition of the reinforcement, Ni60. Solution hardening in γ-Ni and dispersion strengthening of the hard interstitial compounds, such as Cr_7C_3, CrB, Cr_2B, and $Cr_{23}C_6$ increased the hardness of Ni60, and then the HEA/Ni60 coating.

During wear testing at 25 °C, adhesive wear and abrasive wear occurred, while at 500 °C, abrasive wear took place. Volume wear rates of the coating at 25 °C and 500 °C were $0.55 \pm 0.06 \times 10^{-4}$ $mm^3 \cdot N^{-1} \cdot m^{-1}$ and $0.66 \pm 0.02 \times 10^{-4}$ $mm^3 \cdot N^{-1} \cdot m^{-1}$, respectively. Wear resistance of this coating was better than that of the HEA coating, which can result from the addition of the hard Ni60. Therefore, Ni60 is an appropriate reinforcement to further enhance the wear resistance of HEA coating.

Acknowledgments: This work was supported by the National Natural Science Foundation of China (grant number 51401091); a project funded by the Priority Academic Program Development of Jiangsu Higher Education Institutions (PAPD); and a Top-notch Academic Programs Project of Jiangsu Higher Education Institutions (TAPP).

Author Contributions: Lihui Tian and Wei Xiong conceived and designed the experiments; Lihui Tian, Zongkang Feng, and Wei Xiong performed the experiments and analyzed the data; and Lihui Tian wrote the paper.

Conflicts of Interest: The authors declare no conflict of interest.

References

1. Cantor, B.; Chang, I.T.H.; Knight, P.; Vincent, A.J.B. Microstructural development in equiatomic multicomponent alloys. *Mater. Sci. Eng. A* **2004**, *375*, 213–218. [CrossRef]
2. Chen, Y.Y.; Duval, T.; Hung, U.D.; Yeh, J.W.; Shih, H.C. Microstructure and electrochemical properties of high entropy alloys—A comparison with type-304 stainless steel. *Corros. Sci.* **2005**, *47*, 2257–2279. [CrossRef]
3. Niu, S.Z.; Kou, H.C.; Guo, T.; Zhang, Y.; Wang, J.; Li, J.S. Strengthening of nanoprecipitations in an annealed $Al_{0.5}$CoCrFeNi high entropy alloy. *Mater. Sci. Eng. A* **2016**, *671*, 82–86. [CrossRef]

4. Zhang, K.B.; Fu, Z.Y.; Zhang, J.Y.; Wang, W.M.; Lee, S.W.; Niihara, K. Characterization of nanocrystalline CoCrFeNiTiAl high-entropy solid solution processed by mechanical alloying. *J. Alloys Compd.* **2010**, *495*, 33–38. [CrossRef]
5. Peng, Z.; Liu, N.; Zhang, S.Y.; Wu, P.H.; Wang, X.J. Liquid-phase separation of immiscible $CrCu_xFeMo_yNi$ high-entropy alloys. *Mater. Sci. Technol.* **2017**, *33*, 1352–1359. [CrossRef]
6. Wu, P.H.; Peng, Z.; Liu, N.; Niu, M.Y.; Zhu, Z.X.; Wang, X.J. The effect of Mn content on the microstructure and properties of $CoCrCu_{0.1}Fe_{0.15}Mo_{1.5}Mn_xNi$ near equiatomic alloys. *Mater. Trans. JIM* **2016**, *57*, 5–8. [CrossRef]
7. Soare, V.; Burada, M.; Constantin, I.; Mitrică, D.; Bădiliţă, V.; Caragea, A.; Târcolea, M. Electrochemical deposition and microstructural characterization of AlCrFeMnNi and AlCrCuFeMnNi high entropy alloy thin films. *Appl. Surf. Sci.* **2015**, *358*, 533–539. [CrossRef]
8. Braeckman, B.R.; Boydens, F.; Hidalgo, H.; Dutheil, P.; Jullien, M.; Thomann, A.L.; Depla, D. High entropy alloy thin films deposited by magnetron sputtering of powder targets. *Thin Solid Films* **2015**, *580*, 71–76. [CrossRef]
9. Xu, J.H.; Chen, J.; Yu, L.H. Influence of Si content on the microstructure and mechanical properties of VSiN films deposited by reactive magnetron sputtering. *Vacuum* **2016**, *131*, 51–57. [CrossRef]
10. Zhang, H.; Pan, Y.; He, Y.Z.; Jiao, H.S. Microstructure and properties of 6FeNiCoSiCrAlTi high-entropy alloy coating prepared by laser cladding. *Appl. Surf. Sci.* **2011**, *257*, 2259–2263. [CrossRef]
11. Jin, Y.J.; Li, R.F.; Zheng, Q.C.; Li, H.; Wu, M.F. Structure and properties of laser-cladded Ni-based amorphous composite coatings. *Mater. Sci. Technol.* **2016**, *32*, 1206–1211. [CrossRef]
12. Ang, A.S.M.; Berndt, C.C.; Sesso, M.L.; Anupam, A.; Praveen, S.; Kottada, R.S.; Murty, B.S. Plasma-sprayed high entropy alloys: Microstructure and properties of AlCoCrFeNi and MnCoCrFeNi. *Metall. Mater. Trans. A* **2015**, *46*, 791–800. [CrossRef]
13. Wang, L.M.; Chen, C.C.; Yeh, J.W.; Ke, S.T. The microstructure and strengthening mechanism of thermal spray coating $Ni_xCo_{0.6}Fe_{0.2}Cr_ySi_zAlTi_{0.2}$ high-entropy alloys. *Mater. Chem. Phys.* **2011**, *126*, 880–885. [CrossRef]
14. Tian, L.H.; Xiong, W.; Liu, C.; Lu, S.; Fu, M. Microstructure and wear behavior of atmospheric plasma-sprayed AlCoCrFeNiTi high-entropy alloy coating. *J. Mater. Eng. Perform.* **2016**, *25*, 5513–5521. [CrossRef]
15. Oñoro, J. High-temperature mechanical properties of aluminium alloys reinforced with titanium diboride (TiB_2) particles. *Rare Met.* **2011**, *499*, 421–426. [CrossRef]
16. Shao, J.Z.; Li, J.; Song, R.; Bai, L.L.; Chen, J.L.; Qu, C.C. Microstructure and wear behaviors of TiB/TiC reinforced Ti_2Ni/α(Ti) matrix coating produced by laser cladding. *Rare Met.* **2016**. [CrossRef]
17. Simunovic, K.; Saric, T.; Simunovic, G. Different approaches to the investigation and testing of the Ni-based self-fluxing alloy coatings—A review. Part 1: General facts, wear and corrosion investigations. *Tribol. Trans.* **2014**, *57*, 955–979. [CrossRef]
18. Simunovic, K.; Saric, T.; Simunovic, G. Different approaches to the investigation and testing of the Ni-based self-fluxing alloy coatings—A review. Part 2: Microstructure, adhesive strength, cracking behavior, and residual stresses investigations. *Tribol. Trans.* **2014**, *57*, 980–1000. [CrossRef]
19. Natarajan, S.; Anand, E.E.; Akhilesh, K.S.; Rajagopal, A.; Nambiar, P.P. Effect of graphite addition on the microstructure, hardness and abrasive wear behavior of plasma sprayed NiCrBSi coatings. *Mater. Chem. Phys.* **2016**, *175*, 100–106. [CrossRef]
20. *ISO 14916 Thermal Spraying–Determination of Tensile Adhesive Strength*; ISO: Geneva, Switzerland, 1999.
21. Tian, L.H.; Fu, M.; Xiong, W. Microstructural evolution of AlCoCrFeNiSi high-entropy alloy powder during mechanical alloying and its coating performance. *Materials* **2018**, *11*, 320. [CrossRef] [PubMed]
22. Houdková, Š.; Smazalová, E.; Vostřák, M.; Schubert, J. Properties of NiCrBSi coating, as sprayed and remelted by different technologies. *Surf. Coat. Technol.* **2014**, *253*, 14–26. [CrossRef]
23. McPherson, R. The relationship between the mechanism of formation, microstructure and properties of plasma-sprayed coatings. *Thin Solid Films* **1981**, *83*, 297–310. [CrossRef]
24. Li, C.J.; Ohmori, A.; McPherson, R. The relationship between microstructure and Young's modulus of thermally sprayed ceramic coatings. *J. Mater. Sci.* **1997**, *32*, 997–1004. [CrossRef]
25. Li, C.J.; Ohmori, A. Relationships between the microstructure and properties of thermally sprayed deposits. *J. Therm. Spray Technol.* **2002**, *11*, 365–374. [CrossRef]

26. Xing, Y.Z.; Li, C.J.; Li, C.X.; Yang, G.J. Influence of through-lamella grain growth on ionic conductivity of plasma-sprayed yttria-stabilized zirconia as an electrolyte in solid oxide fuel cells. *J. Power Sources* **2008**, *176*, 31–38. [CrossRef]
27. Tung, C.C.; Yeh, J.W.; Shun, T.T.; Chen, S.K.; Huang, Y.S.; Chen, H.C. On the elemental effect of AlCoCrCuFeNi high-entropy alloy system. *Mater. Lett.* **2007**, *61*, 1–5. [CrossRef]
28. Lin, C.W.; Tsai, M.H.; Tsai, C.W.; Yeh, J.W.; Chen, S.K. Microstructure and aging behaviour of $Al_5Cr_{32}Fe_{35}Ni_{22}Ti_6$ high entropy alloy. *Mater. Sci. Technol.* **2015**, *31*, 1165–1170. [CrossRef]
29. Yu, Y.; Liu, W.M.; Zhang, T.B.; Li, J.S.; Wang, J.; Kou, H.C.; Li, J. Microstructure and tribological properties of $AlCoCrFeNiTi_{0.5}$ high-entropy alloy in hydrogen peroxide solution. *Metall. Mater. Trans. A* **2014**, *45*, 201–207. [CrossRef]
30. Leitner, J.; Dubsky, J.; Had, J.; Hanousek, F.; Kolman, B.; Volenik, K. Metastable chromium-rich oxide formed during plasma spraying of high-alloy steel. *Oxid. Met.* **2000**, *54*, 549–558. [CrossRef]
31. Pawlowski, L. *The Science and Engineering of Thermal Spray Coatings*, 2nd ed.; John Wiley & Sons: New York, NY, USA, 2008.
32. Li, J.F.; Ding, C.X. Study on Vickers hardness of plasma sprayed Cr_3C_2-NiCr coating. *J. Chin. Ceram. Soc.* **2000**, *28*, 223–228.
33. Luo, Q. Temperature dependent friction and wear of magnetron sputtered coating TiAlN/VN. *Wear* **2011**, *271*, 2058–2066. [CrossRef]

Article

Research on Microstructure, Mechanical and Tribological Properties of Cr-Ti-B-N Films

Lihua Yu *, Huang Luo, Jianguo Bian, Hongbo Ju and Junhua Xu

School of Materials Science and Engineering, Jiangsu University of Science and Technology, Mengxi Road 2, Zhenjiang 212003, China; daniferlh@163.com (H.L.); joycexql@163.com (J.B.); hbju@rocketmail.com (H.J.); jhxu@just.edu.cn (J.X.)

* Correspondence: lhyu6@just.edu.cn; Tel.: +86-150-5111-9966

Received: 30 June 2017; Accepted: 29 August 2017; Published: 3 September 2017

Abstract: In this study, the Cr-Ti-B-N composite thin films with different Ti and B contents in the films were fabricated by a reactive magnetron sputtering system using chromium (Cr) and TiB_2 targets. The microstructure, mechanical properties, and friction properties over broad ranges of temperature were detected by XRD, SEM, TEM, nano-indentation, 3D profilometer, and tribometer. It was found that the Cr-Ti-B-N films exhibited a double face-centered cubic (fcc)-CrN and amorphous-BN structure. As the Ti and B elements were introduced, the hardness and elastic modulus of Cr-Ti-B-N films increased from 21 and 246.5 GPa to a maximum value of approximately 28 and 283.6 GPa for $Cr_{38.7}Ti_{3.7}B_{6.4}N_{51.2}$ film, and then decreased slightly with a further increase of Ti and B contents. The hardness enhancement was attributed to the residual compressive stress. The friction and wear resistance of the film was improved obviously by the addition of Cr and B, as compared with the CrN film under 200 °C.

Keywords: Cr-Ti-B-N films; magnetron sputtering; microstructure; friction and wear

1. Introduction

In recent years, research in microelectronics mechanical systems, especially silicon base MEMS (Micro Electro Mechanical Systems), has been faced with reliability and service life problems. This is as a result of friction between surfaces such as the shafts, sleeves, pipes, gears, and the micro motor, which is in relative motion and also wears of parts under light load.

Nanometer super-hard film provides a new and effective solution for micro mechanical abrasion problems and is likely to be used as the next generation of sliding, speaking, reading, and writing disk protective film, it can also be used for print head of thermal printer, high fax machine head, and contact sensors because of its wear resistance properties.

Since the 1980s, mechanical properties of CrN_x coatings had been studied, and Cr_2N demonstrated predominant higher hardness and CrN were provided with better oxidation resistance and lower friction coefficient [1]. Due to its excellent properties such as low friction coefficient, high corrosion resistance, wear resistance, high toughness, and low internal stress, transition metal nitrides CrN_x have been widely used as protective hard films to increase the lifetime and performance of cutting and wear proof tools today. While the main drawback of its hardness (20 GPa) [2] restrict the use of CrN. A new generation of intelligent film, which can function very well under extreme conditions, such as high speed, dry cutting, high temperature, and surface oxidation is thus needed. Therefore, the study of friction and wear behavior of thin film over broad range of ambient conditions is of great significance. It has been reported that soft phases (Ag, Cu, Au, etc.) were effective to decrease the friction coefficient over broad working temperature range. Lihua Yu et al. [3] synthesized the WCN-Ag film containing 6.6 at.% Ag has a low friction coefficient (<0.4) in the temperature range of 25–600 °C. The NbN-Ag composite films [4] reported a decreased friction coefficient from 0.65 at 0 at.% Ag to 0.35

at 19.9 at.% Ag at room temperature. Shtansky et al. [5] deposited MoCN-Ag films which maintained low friction coefficient over a wide temperature. Similarly, the role of V element working at high temperature as a result of the generation of V_2O_5 lubricating phase also has been studied extensively, such as CrV_xN [6], TiAlVN [7], TiAlN/VN [8], NbVSiN [9] etc. To further improve the properties of CrN, alloying with other element has been explored extensively [10–12]. The superhardness of TiB_2 films has been detected [13–15]. Many researchers studied the properties of Ti-Cr-B-N film [16,17]. However, there are few data on evaluating the performance of Cr based Cr-Ti-B-N thin film and no reports about the tribological properties at elevated temperature is available as of now.

In this work, the effects of the addition of Ti and B elements on the microstructure, mechanical, and tribological properties of such films were thoroughly investigated.

2. Experimental Details

Cr-Ti-B-N films with different Ti and B content were deposited on silicon wafer and AISI304 stainless steel plates (15 mm × 15 mm × 2 mm) using radio frequency reactive magnetron sputtering system with a pure Cr (99.95 at.%) and a TiB_2 compound target (99.9 at.%). Prior to films deposition, the substrates (stainless steel AISI304) were polished with sandpaper and soft cloth used to clean the surface to removed residual scratches. The polished substrate was cleaned for 15 min ultrasonically in acetone and alcohol, then sent into the sputtering chamber. The base pressure of sputtering chamber was reduced to less than 6.0×10^{-4} Pa, and the pressure during deposition was set at 3.0×10^{-1} Pa. The distance between targets and substrates was 78 mm. A two hour deposition process was carried out in the presence of an Argon gas (Ar) and Nitrogen gas (N_2) with constant Chromium target power and various TiB_2 target power. Before the sputtering of Cr-Ti-B-N films, 10 min deposition of Cr (target power, 200 W) was conducted in a pure Argon atmosphere as the transition layer to confirm a good adhesion force of the films. The samples were located on a rotating plate which ensures homogeneous and uniform films deposited on the surface of the substrates.

The phase structures were evaluated with glancing incident X-ray diffraction with a Siemens X-ray diffractometer, with Cu Kα radiation, operated at 30 kV and 40 mA. The cross-sectional morphologies and component analysis of thin films were examined with a field emission scanning electron microscopy (FE-SEM, Merlin Compact-6170, ZEISS, Jena, Germany) which was operated at 20 kV and an Energy Disperse Spectroscopy (EDS Merlin Compact-6170, ZEISS, Germany) respectively.

The microstructure was analyzed by a high resolution transmission electron microscopy (HRTEM, JEOL, JEM-2100F, Tokyo, Japan) operated at 200 kV. The nanoindentation hardness and elastic modulus were all investigated by means of nanoindenter (CPX + NHT^2 + MST), equipped with a diamond Berkovich indenter tip (3-side pyramid). An automatic indentation mode was programmed to place indentations with a 3 × 3 array, averaging data on the hardness of the film. The maximum load of 3 mN was used to meet the criterion of $d/h < 0.1$, averting the influence of the substrate during the hardness test, where d is the indenter penetration depth into the film, h is the film thickness. The ball-on-disc dry wear test was conducted on the stainless steel substrates at ambient conditions by using a UTM-2 CETR tribometer with an alumina ball as counterpart (diameter, 9.38 mm). The normal load, sliding speed, sliding time, and radius of wear track during the tests were 3 N, 50 r/min, 30 min and 4 mm, respectively. The wear tracks morphologies were examined with a 3D profilometer (Bruker DEKTAK-XT, Billerica, MA, USA) and wear rate (W_s) of the films were calculated using Equation (1).

$$W_s = \frac{C \times S}{F \times L} \tag{1}$$

where C is the perimeter of the wear crack (mm), S is the wear crack area (mm^2), F is the normal load (N) and L is the sliding length. The residual stress of the film was determined by means of the Stoney's formula (Equation (2)).

$$\sigma = \frac{E_s T_s^2 (\frac{1}{R_s} - \frac{1}{R_f})}{6(1 - \nu_s) T_f} \tag{2}$$

where E_s is Young's modulus, T_s is thickness of substrate, R_s is curvature radius of substrate, ν_s is Poisson's ratio of silicon substrate, R_f is curvature radius of film, and T_f is thickness of the film. The curvature radii were also examined by the Bruker 3D Profiler.

3. Results and Discussion

3.1. Elemental Compositions and Microstructure of Cr-Ti-B-N Film

Table 1 shows the elemental compositions and film thickness of Cr-Ti-B-N films with various the powers of TiB_2 composite target. As shown in Table 1, with the increase of the power of TiB_2 composite target, the Cr content in the films gradually decreases, with a corresponding increase of B and Ti content. In addition, the film thickness of the films increases with the increase of the power of TiB_2 composite target.

Table 1. Elemental compositions and film thickness of Cr-Ti-B-N films with various the powers of TiB_2 composite target.

TiB_2 Target Power (W)	Cr (at.%)	Ti (at.%)	B (at.%)	N (at.%)	Thickness of Film (μm)
0	53.7 ± 2.7	0	0	46.3 ± 2.4	1.1 ± 0.1
80	48.6 ± 2.4	1.7 ± 0.1	2.4 ± 0.1	47.3 ± 2.4	1.2 ± 0.1
100	43.9 ± 2.2	2.4 ± 0.1	4.0 ± 0.2	49.7 ± 2.5	1.2 ± 0.1
120	38.7 ± 1.9	3.7 ± 0.2	6.4 ± 0.3	51.2 ± 2.6	1.3 ± 0.1
140	33.4 ± 1. 7	5.3 ± 0.3	7.6 ± 0.4	53.7 ± 2.7	1.4 ± 0.1

Figure 1 shows the X-ray diffraction patterns of Cr-Ti-B-N films with various Ti and B contents. As shown in Figure 1, for the binary CrN film, three diffraction peaks at ~36°, ~44° and ~78° are detected, which are corresponding to the face-centered cubic (fcc) CrN (111), Cr interlayer and fcc-CrN (311) repectively, based on the ICCD card 11-0065. For each pattern of quaternary Cr-Ti-B-N films, the diffraction peaks of the films are similar to that of the binary CrN film, and no other diffraction peaks corresponding to the crystalline phases such as TiB_2 and CrB_2 are not observed. Besides this, the diffraction peaks of Cr-Ti-B-N films gradually shift to higher angle, this suggests that the combination of Ti and B elements into the Cr-Ti-B-N films leads to the decrease of the lattice constant. The grain size of the films was calculated by the Scherrer's equation and the results show that the grain size of the films decreases gradually from about 33 nm for binary CrN film to 15 nm for quaternary $Cr_{33.4}Ti_{5.3}B_{7.6}N_{53.7}$ film.

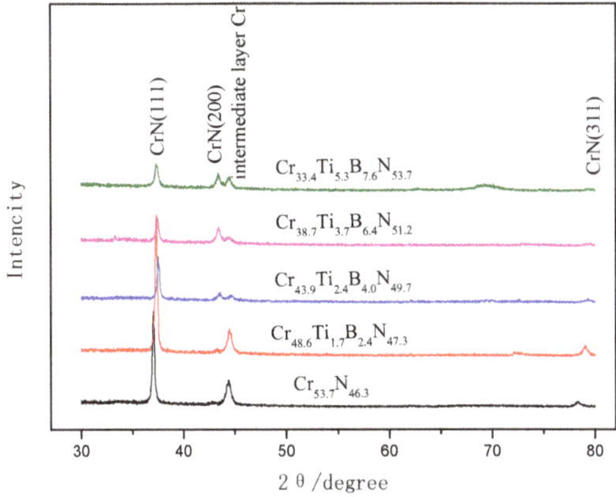

Figure 1. X-ray diffraction patterns of Cr-Ti-B-N films.

Figure 2 illustrates the cross-sectional SEM images of Cr-Ti-B-N films with various Ti and B contents. As illustrated in the Figure 2a, the binary CrN film exhibits a columnar-growth and dense structure. As the increase of the Ti and B contents in the films, as shown in Figure 2b, the columnar-growth and dense structure disappears and the $Cr_{33.4}Ti_{5.3}B_{7.6}N_{53.7}$ film shows an isometric structure.

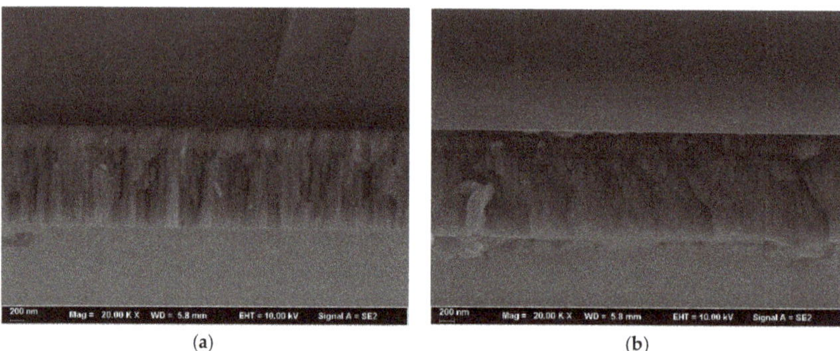

(**a**) (**b**)

Figure 2. The cross-sectional SEM images of Cr-Ti-B-N films with various Ti and B contents: (**a**) CrN; (**b**) $Cr_{33.4}Ti_{5.3}B_{7.6}N_{53.7}$.

In order to further analyze the microstructure of Cr-Ti-B-N films, the microstructure of $Cr_{43.9}Ti_{2.4}B_{4.0}N_{49.7}$ film is investigated by high-resolution transmission electron microscope (HRTEM) and the results is shown in Figure 3. As shown in Figure 3, the $Cr_{43.9}Ti_{2.4}B_{4.0}N_{49.7}$ film exhibits a nano-composite microstructure and is consisted of nanocrystallites and amorphous phase. Besides this, three different lattice fringes with the lattice spacing of ~0.240 nm, ~0.204 nm, and ~0.156 nm are detected. Those lattice fringes are corresponding to the fcc-CrN (111), (200), and (220), since the values of lattice spacing of fcc-CrN (111), (200), and (220) (ICCD card 11-0065) are 0.2394 nm, 0.2068 nm, 0.1463 nm, and 0.1249 nm, respectively. The corresponding selected area electron diffraction (SEAD) pattern of $Cr_{43.9}Ti_{2.4}B_{4.0}N_{49.7}$ film is also shown in Figure 3 as an inset. The diffraction rings can be referred to as the lattice planes of fcc-CrN (111), (200), (220), and (311). These results are in agreement with the XRD pattern. Similar studies [18] also reported the existence of BN amorphous phase.

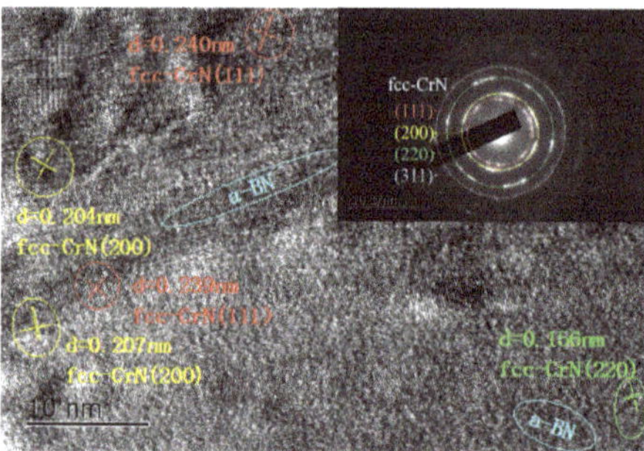

Figure 3. HRTEM micrograph and SAED pattern of $Cr_{43.9}Ti_{2.4}B_{4.0}N_{49.7}$ film.

3.2. Mechanical Properties

Figure 4 shows the hardness and elastic modulus of Cr-Ti-B-N films with the increasing Ti and B contents. As the Ti and B elements were introduced, the hardness and elastic modulus of Cr-Ti-B-N films increased from 21 and 246.5 GPa to a maximum value of approximately 28 and 283.6 GPa for $Cr_{38.7}Ti_{3.7}B_{6.4}N_{51.2}$ film, and then decreased slightly with further increase of Ti and B contents.

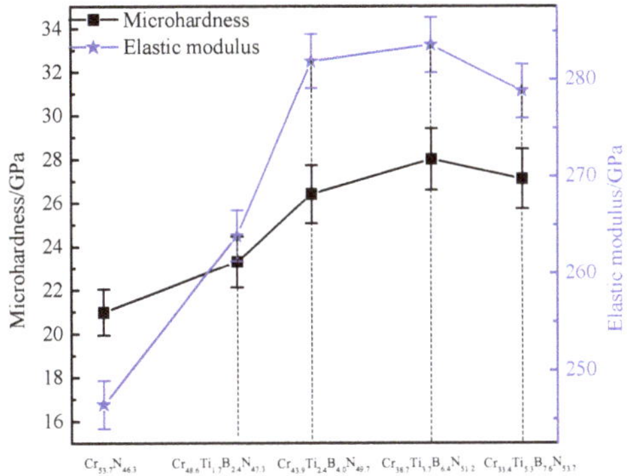

Figure 4. Hardness and elastic modulus of Cr-Ti-B-N films.

The hardness enhancement might be attributed to the solution of Ti into the films due to the different of radius of Ti, Cr atoms. The radius Cr and Ti is 0.125 nm and 0.146 nm, respectively, with the replacement of Ti into the CrN film, a strong lattice distortion will emerge, thus generate large distortion energy. On the other way, CrN and TiN have the same crystal structure with a face-centered cubic lattice type of NaCl. According to the laws of Hume-Rothery, the smaller radius anion (N) occupied face-centered cubic lattice, while Cr and Ti cations were forced into the pores of octahedral; together with the similar crystal lattice parameters between CrN and TiN, it's easy to form a continuous supersaturated solid solution. It can be concluded that the solid solution strengthening, the refined crystalline strengthening, and the increase of residual compressive stress all played a role in enhancing the hardness of the film [19]. While the reduction in hardness with a further increase in Ti and B contents after maximum hardness is as a result of increase of amorphous BN phase [20–22].

Musil [23] and Leyland [24] suggested that the fracture toughness of nanocrystalline films would be improved by high ratios of H/E (resistance against elastic strain to failure) and H^3/E^{*2} (resistance against plastic deformation). By examining the H/E and H^3/E^{*2} values of the films, as shown in Figure 5, it can be seen that Cr-Ti-B-N films exhibit improved H^3/E^{*2} values of between 0.16 and 0.22 GPa, as compared to a low value of 0.13 GPa of CrN film. The variation tendency of the values with H/E and H^3/E^{*2} are corresponding to the variation of the hardness. The highest H/E value of 0.098 was found in $Cr_{38.7}Ti_{3.7}B_{6.4}N_{51.2}$ films, suggesting this film may exhibit better toughness and wear resistance as compared to the other films.

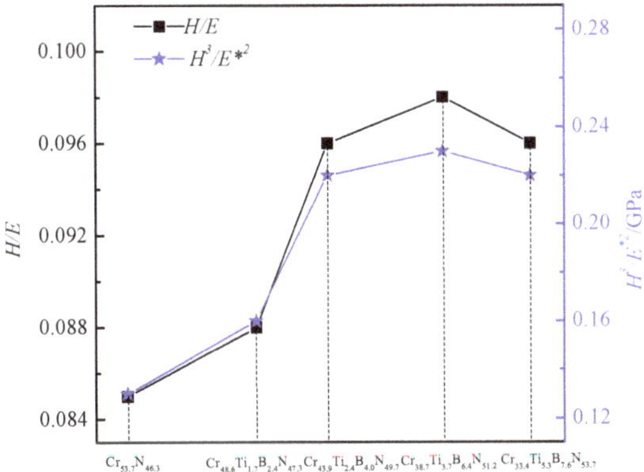

Figure 5. Calculated H/E and H^3/E^{*2} values of Cr-Ti-B-N films.

Figure 6a shows the load-displacement curves of $Cr_{38.7}Ti_{3.7}B_{6.4}N_{51.2}$ film (3.7 at.% Ti) obtained from the nanoindentation measurement. The amounts of Plastic deformation and Elastic deformation are shown clearly. Figure 6b shows amount of Plastic deformation and Elastic deformation of Cr-Ti-B-N films with various Ti and B contents. We can conclude that with an increasing of the Ti and B contents, the plastic deformation depths are 50, 41.3, 38.5, 33.6, 36 nm; the elastic deformation displacements are 46.3, 47.6, 48, 49.6, 48.4 nm, respectively which are corresponding to the results of examined H/E and H^3/E^{*2} values. The indentation depth decreased with an increasing of the Ti and B contents whiles the hardness increases with an increase of the Ti and B contents.

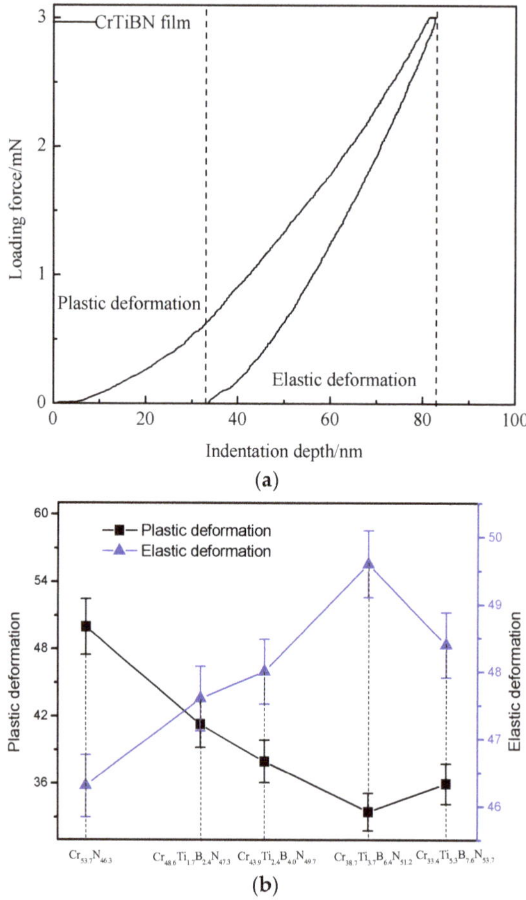

Figure 6. (**a**) Curves of nanoindention of $Cr_{38.7}Ti_{3.7}B_{6.4}N_{51.2}$ film; (**b**) amount of Plastic deformation and Elastic deformation of Cr-Ti-B-N films.

3.3. *Tribological Performance at Room Temperature*

The ball-on-disk wear test was conducted to evaluate the tribological properties of the films. The wear tests on the film with various Ti and B contents using an Alumina (Al_2O_3) ball was performed. As shown in Figure 7, the wear rate and friction coefficient decreased with an increase in Ti and B contents from $Cr_{53.7}N_{46.3}$ film to $Cr_{33.4}Ti_{5.3}B_{7.6}N_{53.7}$ film and a further increases in Ti and B contents resulted in increase in both wear rate and friction coefficient. Hence the optimal parameter was obtained at a Titanium content of 3.7 at.%, which is in complete agreement with the suggestion that higher wear resistance is achieved with better *H/E* and H^3/E^{*2} values. The addition of Ti and B marginally decreased the friction coefficient at room temperature.

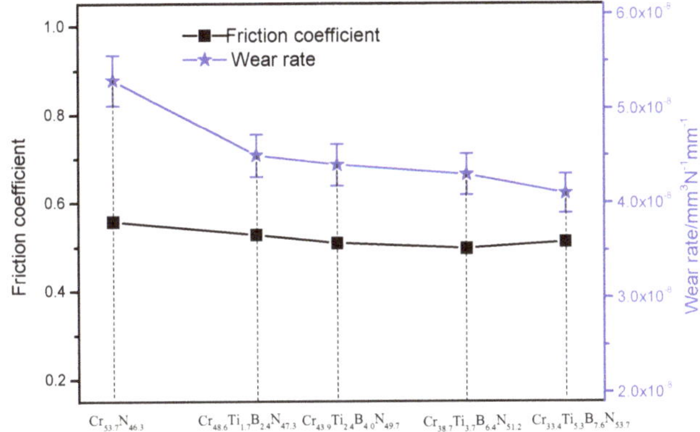

Figure 7. Friction coefficient and wear rate of Cr-Ti-B-N films at room temperature.

The wear track morphologies of Cr-Ti-B-N films with different Ti and B contents are shown in Figure 8. The confined and superficial wear track corresponded with the uppermost hardness and a better H/E value when the Ti content reached 3.7%. From Figure 8 we can conclude that the main wear patterns of Cr-Ti-B-N film with different Ti and B contents was abrasive wear, and the wear debris were accumulated at the edge of the wear track. In general, the film showed excellent wear resistance.

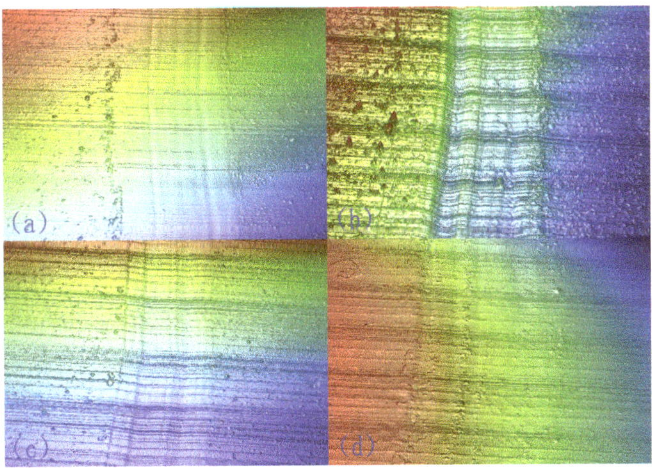

Figure 8. 3D wear track morphologies of Cr-Ti-B-N films: (**a**) $Cr_{48.6}Ti_{1.7}B_{2.4}N_{47.3}$; (**b**) $Cr_{43.9}Ti_{2.4}B_{4.0}N_{49.7}$; (**c**) $Cr_{38.7}Ti_{3.7}B_{6.4}N_{51.2}$ and (**d**) $Cr_{33.4}Ti_{5.3}B_{7.6}N_{53.7}$.

3.4. Tribological Performance at Elevated Temperature

Plots of friction coefficient about CrN film and $Cr_{38.7}Ti_{3.7}B_{6.4}N_{51.2}$ composite film at different temperature are shown in Figure 9. The friction coefficient of both films decreased continually and then tended to a stable value while the value tested at 800 °C showed a decline. The minimum friction coefficient of CrN film was 0.32 obtained at 400 °C while the CrTiBN composite film was 0.38 acquired at 800 °C. As shown in Figure 9, the addition of Ti and B elements significantly improved the friction resistance under 200 °C. This attributed to the production of boron oxide which with lower shear modulus. With the elevated temperature higher than 400 °C, the boron oxide will melt and volatilization gradually and lots of defects generated at grain boundary, resulting in the rise of friction coefficient [25].

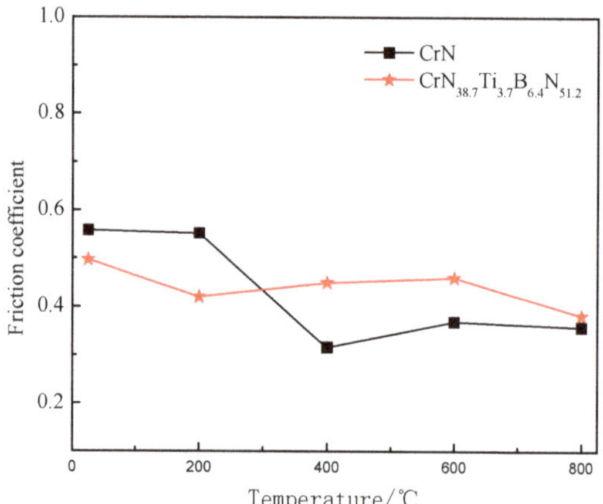

Figure 9. Friction coefficient of CrN film and $Cr_{38.7}Ti_{3.7}B_{6.4}N_{51.2}$ composite film at different temperature.

Figure 10 shows the wear rate of CrN film and $Cr_{38.7}Ti_{3.7}B_{6.4}N_{51.2}$ composite film at different temperatures. It can be observed that the wear rate increases with an increase in temperature. With the lifted temperature, the wear rate increased gradually which, due to the reduction of hardness and oxidation. The wear rate of CrN film tested at 200 °C rose sharply because of the wear out of the film. The wear resistance of $Cr_{38.7}Ti_{3.7}B_{6.4}N_{51.2}$ film was enhanced obviously under 400 °C.

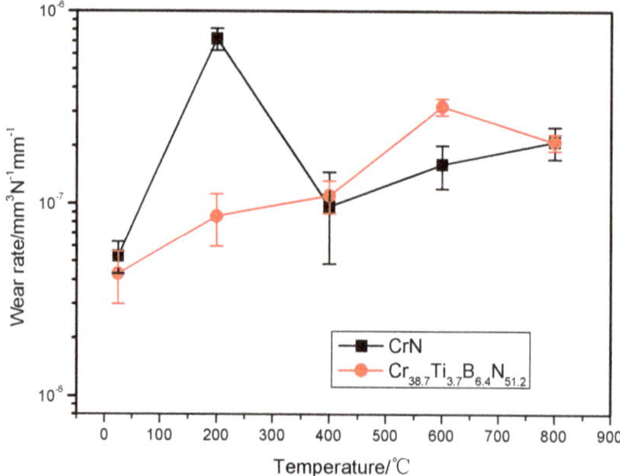

Figure 10. Wear rate of CrN film and $Cr_{38.7}Ti_{3.7}B_{6.4}N_{51.2}$ composite film at different temperature.

Figure 11 shows the friction coefficient of CrN film and $Cr_{38.7}Ti_{3.7}B_{6.4}N_{51.2}$ composite film as a function of wear duration at 200 °C in sliding against Al_2O_3 ball. The CrN film was worn out in 10 min while the $Cr_{38.7}Ti_{3.7}B_{6.4}N_{51.2}$ composite film was worn out after 20 min, which indicated an obvious improvement of the friction and wear resistance ability at 200 °C. The wear out of CrN film leads to a relative high value of the results, which are in good agreement of Figures 9 and 10.

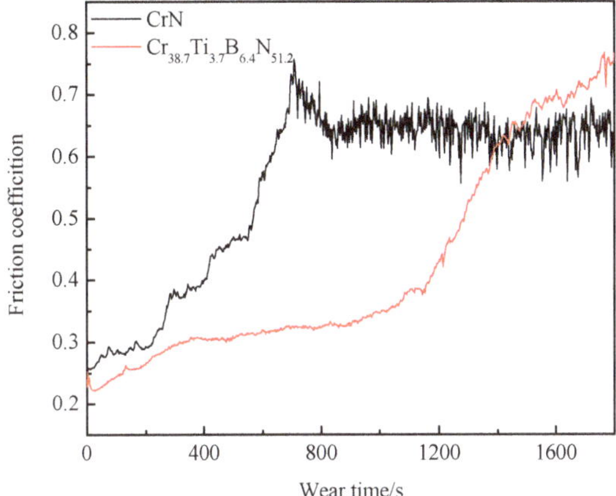

Figure 11. Friction curves of CrN film and $Cr_{38.7}Ti_{3.7}B_{6.4}N_{51.2}$ composite film tested at 200 °C.

To further confirm the changes in wear rate at different temperatures, the XRD pattern of $Cr_{38.7}Ti_{3.7}B_{6.4}N_{51.2}$ film at different temperature are shown in Figure 12. The slight increase in coefficient of friction and wear rate values are attributed to low hardness, the grains growth and the broader cracks at elevated temperature. While the sudden decrease in values at 800 °C may be as a result of the generation of mass lubricants of oxides such as TiO_2, Cr_2O_3 [20,26]. Several oxide peaks of Ti and Cr were detected only at 800 °C which confirms the appearance of oxide lubricants. Pilling and Bedworth first put forward the integrity of the oxide film. The ratio of volume fraction of metal oxide compared to metal is as a judgment of oxide film intactness, namely the PBR (Pilling Bedworth Ratio). Practice has proved that good protective oxide film of PBR value should be $1 < PBR < 2.5$. The PBR values of TiO_2 and Cr_2O_3 were 1.95 and 1.99, respectively. In other words, the temperature of the grinding crack area is higher than the actual set temperature, and then the phase change of materials [27,28] caused by the high temperature and the flash temperature of friction made the wear aggravation.

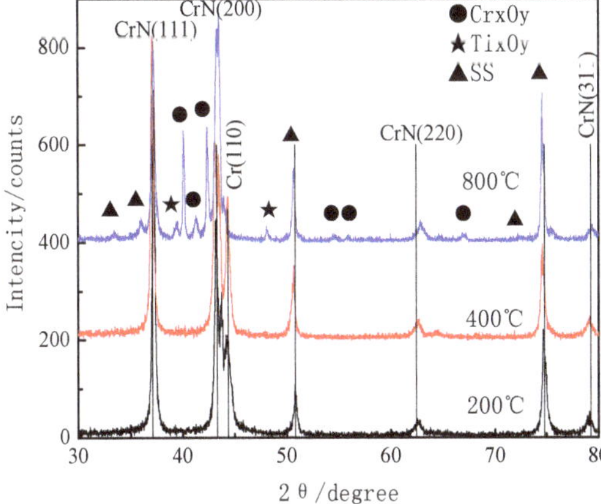

Figure 12. XRD patterns of $Cr_{38.7}Ti_{3.7}B_{6.1}N_{51.2}$ composite film after wear test at different temperature.

Figure 13 shows the changes of atomic fraction of Oxygen (O) of $Cr_{38.7}Ti_{3.7}B_{6.4}N_{51.2}$ film after wear test at different temperature. EDS results show increase in Oxygen content as a function of temperature. Compared with the film out of the grinding crack zone, the oxygen content within the trajectory was significantly increased. This is mainly because friction makes the contact surface temperature higher and then oxidation reactions are more likely to proceed.

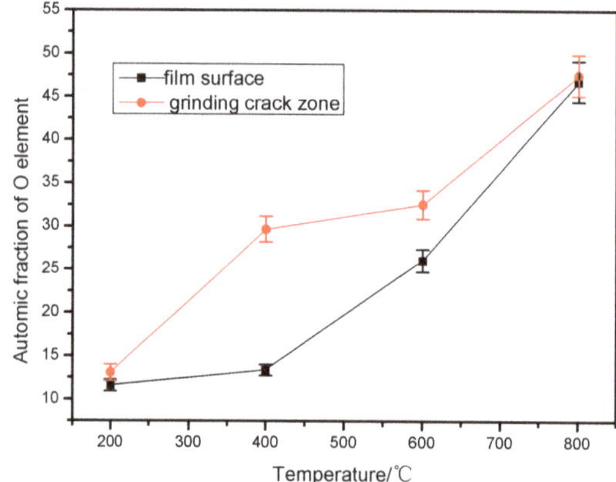

Figure 13. Atomic fraction of O element of the $Cr_{38.7}Ti_{3.7}B_{6.4}N_{51.2}$ film after wear test at different temperature.

The SEM morphologies of wear track of $Cr_{38.7}Ti_{3.7}B_{6.4}N_{51.2}$ film at elevated temperatures are shown in Figure 14. The wear track morphologies at different temperature reveal diverse worn out mechanisms such as abrasive wear, adhesive wear, plastic deformation, and crack etc. The wear track of $Cr_{38.7}Ti_{3.7}B_{6.4}N_{51.2}$ film tested at 800 °C exhibited obvious micro-cracks. Surface micro-cracks may have an effect of storing lubricants and collecting the wear debris produced in the process of wear testing. The hard grinding grain was collected within the micro cracks, reducing the damage for lubricating film. As a result, the friction coefficient and wear rate decreased evidently and thus prolong the life of the film.

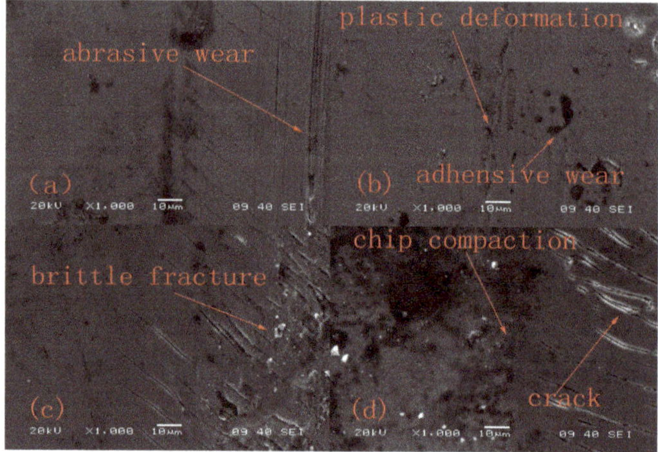

Figure 14. Wear morphology of $Cr_{38.7}Ti_{3.7}B_{6.4}N_{51.2}$ film at elevated temperature characterized by SEM: (**a**) 200 °C; (**b**) 400 °C; (**c**) 600 °C and (**d**) 800 °C.

With the oxidation of materials at different temperatures, there were changes in surface morphology and color of the films; this is due to the diverse morphologies of the film under different temperature and the generation of new oxides, respectively. After the wear test, the color of Cr-Ti-B-N film changed from silver to purple at the temperature of 600 °C; however, the uncoated samples color changed to gold at 400 °C. This indicated a higher oxidation resistance of the Cr-Ti-B-N film.

4. Conclusions

Several films of Cr-Ti-B-N with a pure Cr target at a constant radio frequency power of 200 W and a series of TiB_2 compound targets were fabricated using the reactive magnetron sputtering system. The microstructure, mechanical, and tribological properties of the film was strongly influenced by the TiB_2 target power. The cross-sectional morphology showed that the growth mechanism changed from the columnar crystal to isometric crystal with the addition of Titanium and Boron to CrN film. The Cr-Ti-B-N films exhibited a double face-centered cubic (fcc)-CrN and amorphous-BN structure. As the Ti and B elements were introduced, the hardness and elastic modulus of Cr-Ti-B-N films increased from 21 and 246.5 GPa to a maximum value of approximately 28 and 283.6 GPa for $Cr_{38.7}Ti_{3.7}B_{6.4}N_{51.2}$ film, and then decreased slightly with further increase of Ti and B contents. The hardness enhancement was attributed to the residual compressive stress. The friction and wear resistance of the film was improved obviously by the addition of Cr and B as compared with the CrN film under 200 °C. It can be conclude that the $Cr_{38.7}Ti_{3.7}B_{6.4}N_{51.2}$ film exhibited improved hardness and wear resistance over broad ranges of ambient conditions.

Acknowledgments: This work was supported by National Natural Science Foundation of China (Grants No. 51074080, 51374115 and 51574131).

Author Contributions: Lihua Yu conceived and designed the experiments; Lihua Yu and Huang Luo performed and analyzed the data the experiments; Jianguo Bian, Hongbo Ju and Junhua Xu contributed to data analysis. Lihua Yu wrote the paper; all authors participated and discussed this work and contributed to the submitted and published manuscript.

Conflicts of Interest: The authors declare no conflict of interest.

References

1. Hones, P.; Sanjines, R.; Levy, F. Characterization of sputter-deposited chromium nitride thin films for hard coatings. *Surf. Coat. Technol.* **1997**, *94*, 398–402. [CrossRef]
2. Zeilinger, A.; Daniel, R.; Schöberl, T.; Stefenelli, M.; Sartory, B.; Keckes, J.; Mitterer, C. Resolving depth evolution of microstructure and hardness in sputtered CrN film. *Thin Solid Films* **2015**, *581*, 75–79. [CrossRef]
3. Yu, L.; Zhao, H.; Xu, J. Influence of silver content on structure, mechanical and tribological properties of WCN–Ag films. *Mater. Charact.* **2016**, *114*, 136–145. [CrossRef]
4. Ju, H.; Xu, J. Microstructure and tribological properties of NbN-Ag composite films by reactive magnetron sputtering. *Appl. Surf. Sci.* **2015**, *355*, 878–883. [CrossRef]
5. Shtansky, D.V.; Bondarev, A.V. Structure and tribological properties of MoCN-Ag coatings in the temperature range of 25–700 °C. *Appl. Surf. Sci.* **2013**, *273*, 408–414. [CrossRef]
6. Rapoport, L.; Moshkovich, A.; Perfilyev, V. High temperature friction behavior of CrV_xN coatings. *Surf. Coat. Technol.* **2014**, *238*, 207–215. [CrossRef]
7. Tillmann, W.; Sprute, T.; Hoffmann, F.; Chang, Y.; Tsai, C. Influence of bias voltage on residual stresses and tribological properties of TiAlVN-coatings at elevated temperatures. *Surf. Coat. Technol.* **2013**, *231*, 122–125. [CrossRef]
8. Hovsepian, P.; Lewis, D.; Luo, Q.; Münz, W. TiAlN based nanoscale multilayer coatings designed to adapt their tribological properties at elevated temperatures. *Thin Solid Films* **2005**, *485*, 160–168. [CrossRef]
9. Ju, H.; Xu, J. Influence of vanadium incorporation on the microstructure, mechanical and tribological properties of Nb–V–Si–N films deposited by reactive magnetron sputtering. *Mater. Charact.* **2015**, *107*, 411–418. [CrossRef]

10. Mallia, B.; Stüber, M.; Dearnley, P. Character and chemical-wear response of high alloy austenitic stainless steel (Ortron 90) surface engineered with magnetron sputtered Cr–B–N ternary alloy films. *Thin Solid Films* **2013**, *549*, 216–223. [CrossRef]
11. Lee, J.; Cheng, C.; Chen, H.; Ho, L.; Duh, J.; Chan, Y. The influence of boron contents on the microstructure and mechanical properties of Cr–B–N thin films. *Vacuum* **2013**, *87*, 191–194. [CrossRef]
12. Zhang, G.; Wang, L.; Yan, P.; Xue, Q. Structure and mechanical properties of Cr–B–N films deposited by reactive magnetron sputtering. *J. Alloy. Compd.* **2009**, *486*, 227–232. [CrossRef]
13. Rupa, P.K.P.; Chakraborti, P.C.; Mishra, S.K. Structure and indentation behavior of nanocomposite Ti–B–N films. *Thin Solid Films* **2014**, *564*, 160–169. [CrossRef]
14. Mayrhofer, P.H.; Mitterer, C.; Wen, J.G.; Greene, J.E.; Petrov, I. Self-organized nanocolumnar structure in superhard TiB2 thin films. *Appl. Phys. Lett.* **2005**, *86*, 131909. [CrossRef]
15. Rupa, P.K.P.; Chakraborti, P.C.; Mishra, S.K. Mechanical and deformation behaviour of titanium diboride thin films deposited by magnetron sputtering. *Thin Solid Films* **2009**, *517*, 2912–2919. [CrossRef]
16. Paternoster, C.; Fabrizi, A. Thermal evolution and mechanical properties of hard Ti–Cr–B–N and Ti–Al–Si–B–N coatings. *Surf. Coat. Technol.* **2008**, *203*, 736–740. [CrossRef]
17. Kiryukhantsev-Korneev, P.V.; Shtansky, D.V. Thermal stability and oxidation resistance of Ti–B–N, Ti–Cr–B–N, Ti–Si–B–N and Ti–Al–Si–B–N films. *Surf. Coat. Technol.* **2007**, *201*, 6143–6147. [CrossRef]
18. Ho, L.; Lee, J.; Chen, H. Structure and mechanical property evaluation of Cr–Ti–B–N films. *Thin Solid Films* **2013**, *544*, 380–385. [CrossRef]
19. Ju, H.; Yu, L.; Xu, J. Microstructure, mechanical and trobological properties of TiN-Ag films deposited by reactive magnetron sputtering. *Vacuum* **2017**, *141*, 82–88. [CrossRef]
20. Shtansky, D.V.; Sheveiko, A.N.; Petrzhik, M.I. Hard tribological Ti–B–N, Ti–Cr–B–N, Ti–Si–B–N and Ti–Al–Si–B–N coatings. *Surf. Coat. Technol.* **2005**, *200*, 208–212. [CrossRef]
21. Yu, L.; Zhao, H.; Xu, J. Mechanical, tribological and corrosion performance of WBN composite films deposited by reactive magnetron sputtering. *Appl. Surf. Sci.* **2014**, *315*, 380–386. [CrossRef]
22. Lin, J.; Moore, J.J.; Moerbe, W.C. Structure and properties of selected (Cr–Al–N, TiC–C, Cr–B–N) nanostructured tribological films. *Int. J. Refract. Met. Hard Mater.* **2010**, *28*, 2–14. [CrossRef]
23. Musil, J.; Jirout, M. Toughness of hard nanostructured ceramic thin films. *Surf. Coat. Technol.* **2007**, *201*, 5148–5152. [CrossRef]
24. Leyland, A.; Matthews, A. On the significance of the H/E ratio in wear control: A nanocomposite coating approach to optimised tribological behavior. *Wear* **2000**, *246*, 1–11. [CrossRef]
25. Erdemir, A. A crystal chemical approach to the formulation of self-lubricating nanocomposite coatings. *Surf. Coat. Technol.* **2005**, *200*, 1792–1796. [CrossRef]
26. Ju, H.; He, S.; Yu, L.; Xu, J. The improvement of oxidation resistance, mechanical and tribological properties of W2N films by doping silicon. *Surf. Coat. Technol.* **2017**, *317*, 158–165. [CrossRef]
27. Ju, H.; He, X.; Yu, L.; Xu, J. The microstructure and tribological properties at elevated temperatures of tungsten silicon nitride films. *Surf. Coat. Technol.* **2017**, *326*, 255–263. [CrossRef]
28. Ju, H.; Yu, L.; He, S.; Xu, J. The enhancement of fracture toughness and tribological properties of the titanium nitride films by doping yttrium. *Surf. Coat. Technol.* **2017**, *321*, 57–63. [CrossRef]

Article

A Microstructural and Wear Resistance Study of Stainless Steel-Ag Coatings Produced through Magnetron Sputtering

Claudia L. España P. [1], **Abel A. C. Recco** [2] **and Jhon J. Olaya** [1,*]

[1] Department of Mechanical and Mechatronics Engineering, Universidad Nacional de Colombia, Bogotá 111321, Colombia; clespanap@unal.edu.co

[2] Physics Department, Santa Catarina State University, UDESC, Joinville 89219-710, Brazil; abel.recco@udesc.br

* Correspondence: jjolayaf@unal.edu.co; Tel.: +57-1-316-5000-11208

Received: 16 August 2018; Accepted: 23 October 2018; Published: 26 October 2018

Abstract: This paper presents a study of the tribological properties of stainless steel coatings with varying Ag contents, deposited via magnetron sputtering. The growth of the coatings was done in Ar and Ar + N_2 atmospheres in order to change the crystalline phase in the coating. The analysis of the chemical composition was performed using energy-dispersive X-ray spectroscopy (EDS) and the structural analysis was performed via X-ray diffraction (XRD). The adhesive wear resistance and the friction coefficient were evaluated using the ball-on-disk test with a ball of alumina. The coatings' adhesion was measured with a scratch tester and the mechanical properties were evaluated with a nanoindenter. The morphology of the films and the wear track were characterized via scanning electron microscopy (SEM). By means of XRD, phases corresponding to the body-centred cubic (BCC) structure were found for the coatings deposited in an inert atmosphere and face-centred cubic (FCC) for those deposited in a reactive atmosphere. A more compact morphology was observed in coatings with a higher silver content. The values of the hardness increased with an increase in the silver content and the presence of nitrogen in the coatings. In the wear traces, mainly mechanisms of oxidative and adhesive wear and plastic deformation were found. The coefficient of friction decreased with an increase of silver in the coatings, whereas the wear rate decreased.

Keywords: stainless steel coating; sputtering; wear; adhesion; friction; structure

1. Introduction

Stainless steels have good resistance to corrosion, high resistance to cold work and good formability. One of the stainless steels used in the biomedical, automotive, naval and aeronautical industries is AISI 316L austenitic stainless steel, since it has good corrosion resistance due to the presence of a thin film of chromium oxide on its surface, known as a passive film [1–5]. However, its low mechanical resistance is sometimes not enough for mechanical applications, causing fractures in structural elements due to mechanical stress and fatigue [6,7]. In view of this, it is important to improve the durability of AISI 316L steel by increasing its hardness and wear resistance [8].

The deposition of coatings on steel allows an improvement in its properties such as hardness and resistance to corrosion, wear and fatigue, among others, in comparison with bulk austenitic stainless steels [9–11]. Stainless steel-based coatings deposited by means of the sputtering technique have been characterized in different investigations. It has been found that this type of coating has a chemical composition similar to that of austenitic stainless steel targets; however, a variation in the structure has been found; for instance, a BCC structure is produced for coatings deposited in inert atmospheres (Ar), while for reactive atmospheres (Ar + N_2), coatings are obtained with a mixture of FCC and BCC phases or only FCC, depending on the amount of nitrogen and the substrate temperature used in the process [10,11]. At the same time, in recent years silver (Ag) has become a material widely used as a dopant in thin films, due to its antibacterial activity, hence its use for the purpose of supplying antibacterial properties to biomedical devices. In coatings, silver can form nanoparticles (AgNPs), improving the antimicrobial activity of biomaterials and reducing the coefficient of friction. Among the medical applications of this type of material, coated or integrated with AgNPs to avoid bacterial growth, are wound dressings, contraceptive devices, surgical instruments and external prostheses [12–14]. The deposition of coatings containing silver has been a subject of interest in applications where materials with good tribological and antibacterial properties are required. Evidence has been shown of a tendency to increase the durability of the different types of compounds that contain silver, because in many cases properties such as the wear resistance and the coefficient of friction are improved, which can be attributed to the possible action of silver as a solid lubricant on the surface of the coatings, thus reducing wear [15–17].

We investigated the effect of silver on the microstructure and the mechanical and tribological properties of stainless steel films with varying silver content that were deposited by means of magnetron sputtering on stainless steel substrates. The growth of the coatings was done in Ar and Ar + N_2 atmospheres in order to change the crystalline phase in the coating.

2. Materials and Methods

Stainless steel coatings were deposited by means of DC magnetron sputtering. The discharge was done with a 316L stainless steel target of 10 cm in diameter and 0.64 cm in thickness, where 1, 2 and 4 pieces of Ag of 4 mm × 4 mm were added to the surface of the target. The target was facing up so that the Ag target sat on top of the SS316 target by gravitational force. The base pressure was 9×10^{-3} Pa and the working pressure was fixed at 0.2 Pa. The deposition gas was composed of argon (purity 99.9%) and nitrogen (purity 99.9%) with flux 1.2 and 0.8 sccm, respectively. The power density was 7 W/cm^2. The substrate temperature was 300 °C and the substrate-target distance was maintained at 6 cm. The coatings were deposited on AISI 316L substrates, which were previously sanded with abrasive paper from 180 to 2000 and polished with cloth and 2.0 µm alumina suspension. To remove alumina particles during the final polishing, an amphoteric surfactant was used. Finally, in order to reduce organic impurities, especially fats, the samples were cleaned with acetone and 2-propanol in ultrasound for 10 min. The thickness of the deposited films was measured by means of a Bruker DektakXT profilometer (Billerica, MA, USA). To perform the thickness measurements, silicon substrates were used, on each of which were placed pieces of silicon in order to generate a step during the deposition of the coatings. The thickness of the deposited thin films was between 3.0 and 4.0 µm, with deposition times of 28 and 32 min for inert and reactive atmospheres, respectively. The coatings were designated as shown in Table 1 according to the amount of Ag pieces used and the atmosphere in which they were deposited.

Table 1. Stainless steel coatings deposited with and without Ag pieces and in inert and reactive atmospheres.

Coating	Silver Pieces	Atmosphere
Ar0Ag	0	Ar
Ar1Ag	1	Ar
Ar4Ag	4	Ar
ArN0Ag	0	Ar + N_2
ArN1Ag	1	Ar + N_2
ArN4Ag	4	Ar + N_2

To study the microstructure, an X'Pert PRO PANalytical grazing incidence X-ray diffraction system (GI-XRD) (Panalytical, Almelo, The Netherlands) was used. The incidence angle was set to

1°, while the swept angle was varied from 35° to 85° in 2θ mode (scanning step of 0.02°), with Cu Kα (1.5418 Å) radiation operating at 45 kV and 40 mA. The grain size of the crystalline phases was determined by the Scherrer formula, as shown in Equation (1).

$$\text{Grain size} = \frac{k\lambda}{B\cos\theta} \tag{1}$$

where k is a constant related to the shape factor (0.94), λ is the X-ray wavelength used in the experiment, B is the peak broadening at half maximum intensity in radians (taking into account the subtraction of the experimental error of the measuring equipment, Bexp) and θ is the Bragg angle.

The percentage of lattice distortion or strain (ε), which is directly proportional to the residual stress, was estimated from the difference between the measured lattice spacing and the bulk parameter lattice spacing. The lattice parameter was calculated from the position of the (110), (200) and (111) Bragg reflections in the XRD spectra. The morphology of the films was studied by means of scanning electron microscopy (SEM, 6701F-JEOL, Tokyo, Japan) and the chemical composition was studied at one spot, via an energy-dispersive X-ray system (Shimadzu EDX-720, Tokyo, Japan), which allowed detecting elements from Na (Z = 11) to U (Z = 90). EDS spectra were collected from different locations on a homogeneous zone, where each single scan area was approximately 10 × 10 mm^2.

Hardness measurements were done with a CTER Nano-Micro nanoindenter (CTER, San Jose, CA, USA). For the hardness test, a Berkovich diamond tip with a radius of 200 nm was used, applying a maximum load of 10 μN in all samples in order to measure hardness (H) and reduced modulus (E_r). To separate the contributions of the substrate properties in the film measurements, the indentation depth was below 10% of film thickness. From the reduced modulus, the elastic modulus of the film was calculated using Equation (2) [18].

$$\frac{1}{E_r} = \frac{(1-\nu_f^2)}{E_f} + \frac{(1-\nu_i^2)}{E_i} \tag{2}$$

where E_r is the reduced modulus, E_i and ν_i are the elastic modulus (1140 GPa) and Poisson's ratio (0.07) of the Berkovich indenter, respectively and E_f and ν_f are the elastic modulus and Poisson's ratio (0.29) of the film, respectively.

The pin-on-disc method was carried out to determine the wear resistance. The test was done using a CETR-UMT-2-110 tribometer (CTER, San Jose, CA, USA) with a stationary pin in air under ambient laboratory conditions. A 6 mm diameter alumina ball was used under a normal load of 1 N and 2 N and a sliding speed of 10 and 50 mm/s for 20 s. In order to obtain a stable coefficient of friction value and determine the wear rate of the coatings, additional tests were carried out under a load of 1 N at a speed of 10 mm/s for 10 min. The wear volume (W_v) of the films was calculated according to ASTM G99-17 [19]. The wear track cross profile was measured at least four points of the wear track with a Dektak 150 profilometer (Bruker, Billerica, MA, USA) in order to obtain an average of the wear track width. After the test, the wear tracks were examined using a Bruker contour GT optical profilometer. The wear products were chemically analysed using energy-dispersive X-ray spectroscopy (EDS). The wear rate (W_s) was calculated according to Archard's equation, Equation (3) [20], where F is the normal load (N) and L is the sliding length (mm). The wear rate is reported in mm^3/Nm.

$$W_s = \frac{W_v}{F \cdot L} \tag{3}$$

The adherence of the coating was measured by means of the scratch test technique with a CSM Instruments model Revetest Xpress. For the test, a progressive load was applied from 1 to 60 N. Then the scratch tracks were characterized by means of optical microscopy in order to measure the critical loads of the coatings.

3. Results

The XRD patterns obtained for the stainless steel coatings deposited in an Ar atmosphere are shown in Figure 1. A body-centred cubic structure can be seen, with mixed orientation, where the peaks were indexed as (110), (200) and (211). The BCC phase is associated with the ferritic structure of stainless steel, commonly obtained when the atmosphere of the sputtering process is composed only of Ar, as has been found by other researchers [9–11]. Furthermore, peaks related to the presence of the FCC structure of silver in Ar1Ag and Ar4Ag coatings can be seen [17,21,22]. In the XRD pattern of the target, a face-centred cubic structure with mixed orientation can be seen, where the peaks were indexed as (111), (200) and (220).

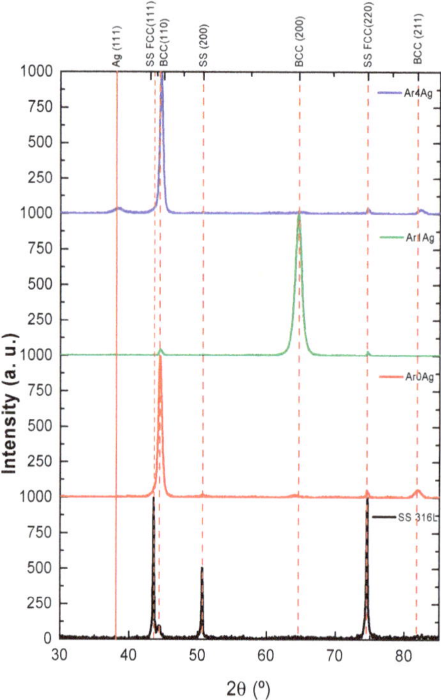

Figure 1. X-ray diffraction (XRD) diffraction patterns of stainless steel coatings deposited in Ar atmosphere.

Figure 2 shows the XRD patterns corresponding to the coatings deposited in the reactive atmosphere (Ar + N_2). A face-centred cubic (FCC) structure can be seen, with a (200) preferential orientation in the sample without Ag (ArN0Ag), because the nitrogen tends to stabilize the FCC structure by broadening the gamma loop in the iron-carbon equilibrium diagram, which has been reported by other researchers [9–11]. However, with the addition of Ag, the (111) signal appears and for the sample with highest Ag content (the ArN4Ag sample) the (111) orientation increases and the (200) disappears. In other studies, it has been reported that the addition of high silver content to different types of materials tends to increase the amorphization of the matrix, which can be attributed to the interruption of the growth of the crystalline phase by the nucleation of Ag nanocrystals at the grain boundaries [22–24].

Table 2 shows the results regarding the size of the crystallite, the percentage of deformation and the lattice parameters. The coatings deposited in a nitrogen atmosphere have a smaller crystallite size and higher values of deformation and lattice parameters. This could be attributed to the inclusion of nitrogen in the octahedral sites of the austenite lattice, generating a lattice distortion as well as expansion of the FCC lattice [9,10,25,26].

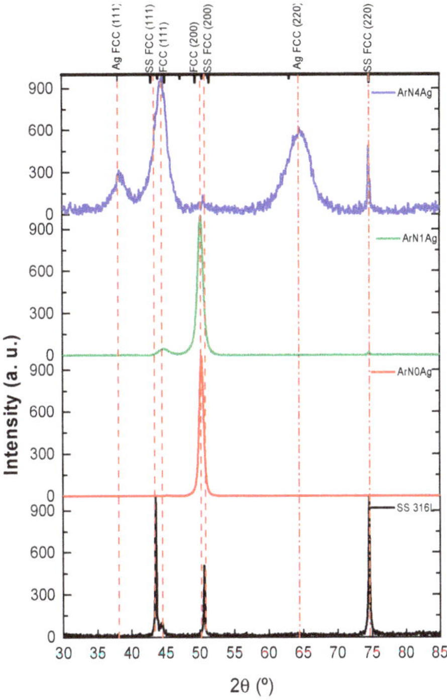

Figure 2. XRD diffraction patterns of stainless steel coatings deposited in Ar + N_2 atmosphere.

Table 2. Results of the size of the crystallite (*D*), percentage of deformation (ε) and lattice parameter (*a*).

Coating	Phase	Reflection (h k l)	*D* (nm)	ε	*a* (nm)
Ar0Ag	BCC	110	15.45	0.10	0.288
Ar1Ag	BCC	200	8.28	0.31	0.289
		110	19.24	0.08	0.288
Ar4Ag	BCC	110	16.46	0.09	0.287
ArN0Ag	FCC	200	12.93	0.15	0.363
ArN1Ag	FCC	200	9.69	0.19	0.363
AgN4Ag	FCC	111	4.24	0.38	0.353

Figure 3 shows the surface morphology and cross section of the stainless steel coating with the highest Ag content (sample Ar4Ag) deposited in an inert atmosphere, where a homogeneous surface, without the presence of defects and a columnar structure that is in accordance with zone I of the Thornton model [27] can be seen. Figure 4 shows the microstructure of the AgN4Ag sample deposited in an atmosphere of nitrogen, in which is seen a more compact morphology with less columnar structure, which could be associated with the T zone structure of the Thornton model [27]. This can be attributed to a combination of the nitrogen in the plasma and the highest perçentage of silver deposited in this coating, which can be segregated at the grain boundaries and columns, interrupting the columnar growth by decreasing the size and number of voids between the columns. Also, the Ag phase is dispersed in the steel crystal, grain growth is inhibited and grain refinement is promoted [28].

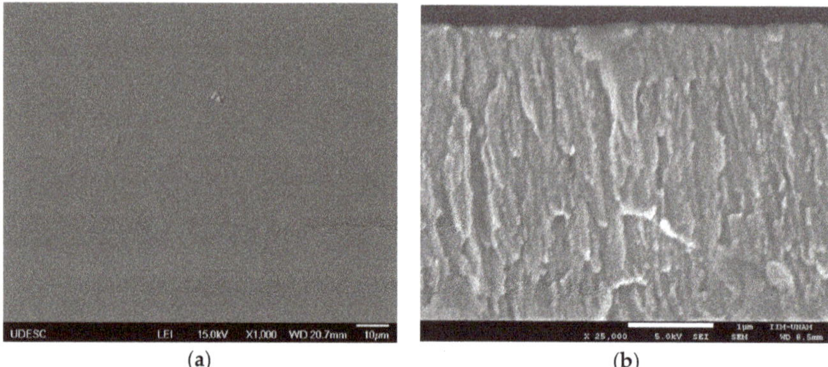

Figure 3. Scanning electron microscopy (SEM) micrographs of Ar4Ag films: (**a**) surface; (**b**) cross-section.

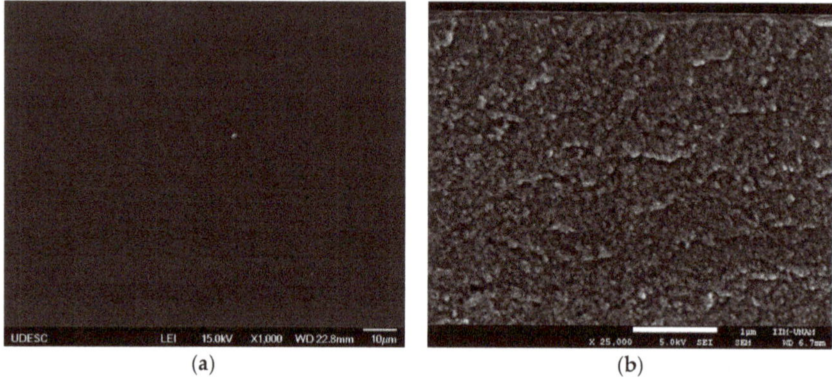

Figure 4. SEM micrographs of ArN4Ag films: (**a**) surface; (**b**) cross-section.

The chemical composition and the mechanic properties are shown in Table 3. The presence of the elements Fe, Cr, Ni, Mn and Mo can be seen, which are commonly found in this type of coating. In addition, there is a slight deviation with respect to the target's chemical composition, due to the difference in sputtering yield between its elements and because the pieces of Ag were added to the surface of the target. Also, an increase in Ag up to 5.25 at.% and 13.06 at.% can be seen in the inert and reactive atmospheres, respectively.

Table 3. Mechanical properties and chemical composition of deposited coatings.

Coating	Hardness (GPa)	Elastic Module (GPa)	H/E	H^3/E^2 (GPa)	Chemical Composition (at.%)					
					Cr	Fe	Ni	Mn	Mo	Ag
Ar0Ag	8.1 ± 0.8	211 ± 14	$0.039 \pm 6.4 \times 10^{-5}$	$0.012 \pm 5.1 \times 10^{-5}$	16.3	69.1	9.6	2.8	2.3	0.0
Ar1Ag	8.6 ± 0.4	210 ± 12	$0.041 \pm 4.2 \times 10^{-5}$	$0.014 \pm 3.6 \times 10^{-5}$	16.3	68.1	9.8	2.6	2.2	1.0
Ar4Ag	10 ± 0.5	242 ± 15	$0.042 \pm 4.7 \times 10^{-5}$	$0.018 \pm 4.9 \times 10^{-5}$	15.7	64.8	9.5	2.6	2.2	5.2
ArN0Ag	8.2 ± 0.3	187 ± 15	$0.044 \pm 5.1 \times 10^{-5}$	$0.016 \pm 4.3 \times 10^{-5}$	16.4	68.6	9.8	3.1	2.2	0.0
ArN1Ag	9.9 ± 0.3	196 ± 8	$0.051 \pm 3.6 \times 10^{-5}$	$0.025 \pm 4.3 \times 10^{-5}$	17.2	67.7	9.5	3.0	2.2	0.5
ArN4Ag	13.2 ± 0.5	214 ± 7	$0.062 \pm 4.4 \times 10^{-5}$	$0.050 \pm 8.9 \times 10^{-5}$	14.4	59.5	8.7	2.4	2.0	13.1
Target 316L	–	–	–	–	17.4	65.6	8.9	3.8	4.2	0.0
Substrate 316L	2.8 ± 0.3	189 ± 15	$0.014 \pm 2.7 \times 10^{-5}$	$0.001 \pm 2.9 \times 10^{-6}$	16.4	69.9	10.1	1.5	2.1	0.0

In Table 3, high values of hardness (values up to 8 GPa) also can be seen in the deposited coatings in comparison with the substrate. However, there was no significant increase with the addition of Ag. It can be seen that the hardness has higher values in the case of coatings produced in a reactive atmosphere, which can be attributed to the fact that the use of nitrogen tends to stabilize the austenitic phase of stainless steels. There was an increase in the lattice parameter and the percentage of deformation due to the supersaturation of nitrogen in the austenitic structure, also known as expanded austenite [10]. Based on the Hall-Petch relation, the hardness can increase with the decrease of the grain size and the increase in hardness is due to the effect of fine grain strengthening. On the other hand, the results of the H/E and H^3/E^2 ratio, where H is the hardness and E the elastic modulus of the coatings, indicate that the coatings with the highest silver content deposited in a nitrogen atmosphere exhibited higher values of these ratios, indicating a highly elastic behaviour [29] and resistance to plastic deformation [30].

Figure 5 shows the track obtained by means of the scratch test technique in coatings deposited in the absence of nitrogen. In general, the coatings have good adherence without the formation of cracks on the track; however, pile-up failure can be seen at the edges of the scratch. Also, a large part of the material is dragged by the indenter and located on the sides of the track, which is associated with stick-slip effects. Plastic deformation phenomena and material removal can be seen, mainly due to micro-cutting mechanisms, with an absence of adhesive or cohesive faults [31,32]. Similar behaviour is seen for the coatings deposited in the presence of nitrogen (see Figure 6). However, in the case of the ArN4Ag coating, there is a severe case of spallation phenomena, which is characterized by large regions of coatings with low adhesion forces, coating detachment, adhesive failure and the absence of cohesive failure [33]. It was determined that the critical load (L_c) at which the failure occurred in this coating was 2.81 N.

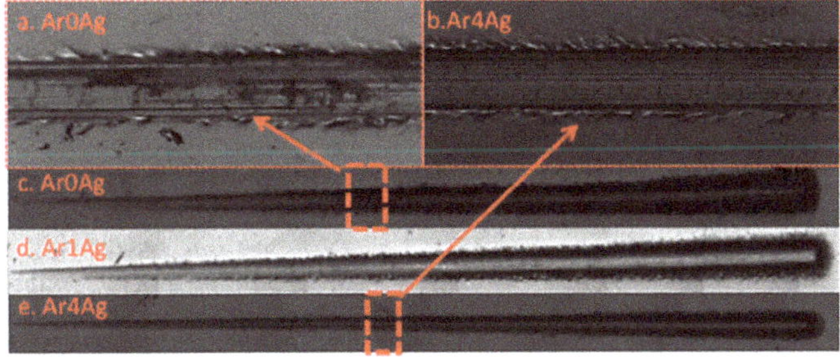

Figure 5. Micrographs of the wear track produced by the scratch tester in coatings deposited in an inert atmosphere: (**a**,**c**) Ar0Ag film; (**b**,**e**) Ar4Ag film; (**d**) Ar1Ag film.

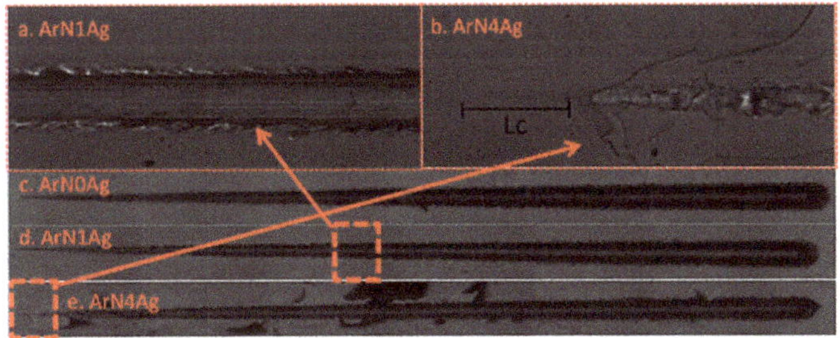

Figure 6. Micrographs of the wear track produced by the scratch tester in coatings deposited in a reactive atmosphere: (**a**,**d**) ArN1Ag; (**b**,**e**) ArN4Ag; (**c**) ArN0Ag.

The coefficient of friction (COF) of the coatings is presented in Figure 7. The coefficient of friction in the produced coatings was reduced in comparison with the substrate. It can be seen that there are no significant changes in the COF value in the coatings deposited in the absence of nitrogen (Figure 7a). As shown, the average COF of the film without Ag was 0.5 but a slight increase in the COF was observed with the change in silver content until it reached 0.54 for the Ar4Ag sample. On the other hand, in the coatings deposited in a reactive atmosphere (Figure 7b), there was a tendency to decrease the COF with an increase in the silver content. As shown, the average COF was 0.5. However, the average COF for Ag films decreases gradually and reaches 0.35 in the ArN4Ag sample. This behaviour can be attributed to a good solid lubricant, Ag, which allows lubricating the wear tracks of the coating and friction pairs, so the sliding between the two surfaces tends to be facilitated, lowering the COF of the film. The quality of silver as a solid lubricant depends on its ability to segregate and hold during a certain number of cycles on the surface, so if the segregation of silver is low, the wear rate will not be affected but if the segregation is high, the wear rate is influenced by the time it takes for the silver to run out, since once exhausted the solid lubrication ability of the film is reduced and the surface has an unstable porous structure, causing failure and an increase in the wear rate [15,16].

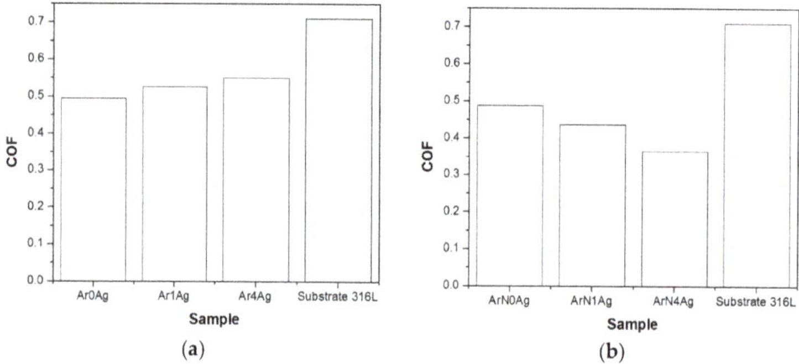

Figure 7. Values of coefficient of friction (COF) of the deposited coatings: (**a**) inert atmosphere; (**b**) reactive atmosphere.

Figure 8 shows the results of the wear rate in the produced coatings. In general, the wear rates in the produced coatings were reduced in comparison with the substrate, which is associated with the significant increase in the hardness of the coatings. Coatings deposited in the presence of nitrogen had lower values of wear rate, possibly attributable to the fact that this group has the highest values of hardness and resistance to plastic deformation. However, it can also be seen that an increase in the silver content in the film produced a higher value of the wear rate, possibly due to an increase in the fragility of the material, loss of adhesion, Ag debris and Ag aggregates formed on the wear tracks. For example, in DLC coatings doped with silver, similar results were found: a decrease in the rate of wear and COF for silver contents between 4.3% and 10.6%. This was attributed to the presence of Ag nanograins, while higher levels of Ag result in a detriment to the tribological properties of the coatings [34,35]. Therefore, Ag aggregates on the wear surface during the sliding process are forming an Ag layer on the counterpart, which may be responsible for the increased wear rate in relation to the reference coating. ZrN/Ag films also showed a decrease in the COF and an increase in the wear rate with a rise in the amount of Ag [28,36]. With an increase in the Ag content in the film, a large amount of the soft Ag phase in the film during the friction experiment is separated from the abrasive surface by the sliding of the friction pair and wear is formed, resulting in the film's wear rate increasing gradually with the Ag content. However, due to the low shear strength of Ag, in the friction process Ag easily breaks away from the wear surface with the friction of the sliding, thus forming debris, resulting in an increase in the wear rate. In addition, more and more Ag diffuses to the surface as the ambient temperature increases, forming voids inside the film which are easily worn under the pressure of the friction pair [28,35].

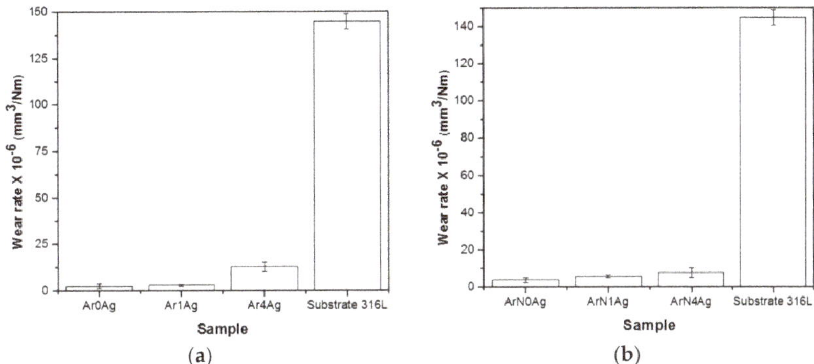

Figure 8. Values of wear rate of the coating deposited: (**a**) inert atmosphere; (**b**) reactive atmosphere.

Figures 9 and 10 show the SEM images of the worn surfaces for samples without Ag and those with higher Ag content and Table 4 shows the chemical composition through EDS of the wear tracks. A mechanism of oxidative wear can be seen for all coatings. This tribo-film is characterized by the presence of oxidized residues on the worn surface, as well as the presence of some scratches and residues of Ag, causing abrasion wear mechanisms [36]. Tribooxidation was a general phenomenon in the wear test at room temperature.

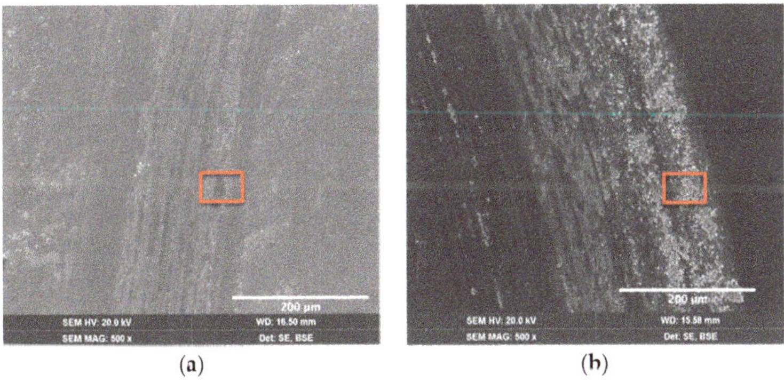

Figure 9. SEM images of wear tracks of coatings deposited in inert atmosphere: (**a**) sample Ar0Ag; (**b**) sample Ar4Ag.

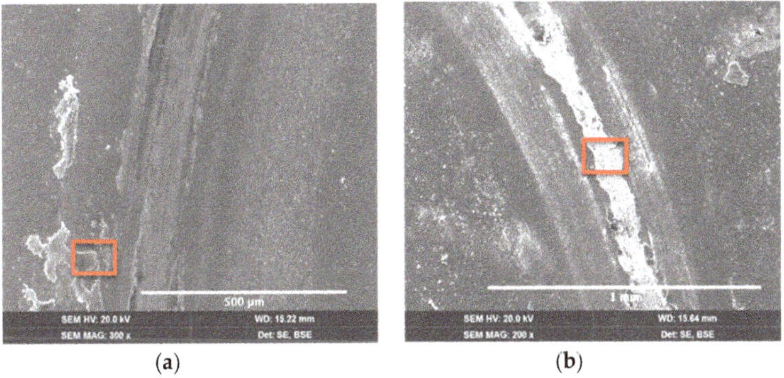

Figure 10. SEM images of wear tracks of coatings deposited in a reactive atmosphere: (**a**) ArN0Ag sample; (**b**) ArN4Ag sample.

Table 4. Results by means of the energy dispersive X-ray spectroscopy (EDS) analyses of wear tracks.

Coating	Chemical Composition (at.%)						
	Fe	Cr	Ni	O	Mn	Al	Ag
Ar0Ag	55.0	13.3	6.3	24.1	0.4	1.0	–
Ar4Ag	40.3	10.1	5.0	42.6	0.3	–	0.3
ArN0Ag	48.2	12.4	6.1	32.8	0.5	–	–
ArN4Ag	32.8	8.5	3.9	52.6	0.3	–	1.9

The coatings deposited in the reactive atmosphere (Figure 10) exhibited greater evidence of oxidative wear, according to the results of the EDS analysis (Table 4), where high percentages of oxygen can be seen in the evaluation of the chemical composition carried out on the wear tracks [25]. The plastic deformation and oxidative wear in these films can generally be attributed to the high concentration of pressure on the surface, causing a local increase in temperature, thus generating oxidation processes during wear [17]. However, delamination occurred in the coatings with a high silver content (Figure 10b). The delamination behaviour of coatings doped with Ag has been reported in other investigations for high silver contents (13.1 at.%) [16,35,36].

4. Conclusions

Stainless steel coatings with varying Ag contents were deposited in inert (Ar) and reactive (Ar and N_2) atmospheres by means of magnetron sputtering and their structural, mechanical and tribological properties were studied. The study concludes the following:

BCC and FCC microstructures were exhibited by the coatings deposited in inert and reactive atmospheres, respectively. As for coatings deposited with silver, the FCC structure of silver was also evident. The coatings deposited in a nitrogen atmosphere and with a higher amount of Ag had a smaller crystallite size and higher values of deformation and lattice parameter.

Nanohardness values up to 8 GPa were obtained for the stainless steel coatings, corresponding to a significant increase in the hardness compared to the hardness of the AISI 316L steel used as substrate. The coatings with Ag and deposited in a reactive atmosphere exhibited a higher degree of hardness, which can be attributed to the fact that the use of nitrogen tends to stabilize the austenitic phase of stainless steel and cause an increase in the lattice parameter and the percentage of deformation, due to the supersaturation of nitrogen in the austenitic structure.

The COF in the produced coatings was reduced in comparison with the substrate. In the coatings deposited in a reactive atmosphere, there was a tendency to decrease the COF with an increase in silver content. This behaviour can be attributed to a good solid lubricant, Ag, which allows lubrication of the wear tracks of the film and friction pairs, so that the sliding between the two surfaces tends to decrease the friction coefficient of the coating.

Finally, the wear rates in the produced coatings were reduced in comparison with the substrate, which is associated with the significant increase in the hardness of the coatings. The coatings deposited with lower Ag content (Ar0Ag and ArN0Ag) have a lower value of the wear rate. The decrease in the wear resistance with the increase of Ag is possibly due to an increase in the fragility of the material, loss of adhesion, Ag aggregates, or debris formed on the wear track.

Author Contributions: C.L.E.P., A.A.C.R. and J.J.O. conceived and designed the experiments; C.L.E.P. performed the experiments; C.L.E.P., A.A.C.R. and J.J.O. wrote the paper.

Funding: This research was funded by National University of Colombia, project "Nanostructured coatings of FCC metallic alloys deposited via sputtering" (37535).

Conflicts of Interest: The authors declare no conflict of interest.

References

1. Antunes, R.A.; Rodas, A.C.D.; Lima, N.B.; Higa, O.Z.; Costa, I. Study of the corrosion resistance and in vitro biocompatibility of PVD TiCN-coated AISI 316 L austenitic stainless steel for orthopedic applications. *Surf. Coat. Technol.* **2010**, *205*, 2074–2081. [CrossRef]
2. Parsapour, A.; Khorasani, S.N.; Fathi, M.H. Effect of surface treatment and metallic coating on corrosion behavior and biocompatibility of surgical 316L stainless steel implant. *J. Mater. Sci. Technol.* **2012**, *28*, 125–131. [CrossRef]
3. Pareja López, A.N.D.R.É.S.; García García, C.P.; Abad Mejía, P.J.; Márquez Fernández, M.E. Citotoxic and genotoxic study of in vitro released products of stainless steel 316L with bioactive ceramic coatings. *Iatreia* **2007**, *20*, 12–20. (In Spanish)
4. Gopi, D.; Ramya, S.; Rajeswari, D.; Kavitha, L. Corrosion protection performance of porous strontium hydroxyapatite coating on polypyrrole coated 316L stainless steel. *Colloids Surf. B* **2013**, *107*, 130–136. [CrossRef] [PubMed]
5. Anghelina, F.V.; Ungureanu, D.N.; Bratu, V.; Popescu, I.N.; Rusanescu, C.O. Fine structure analysis of biocompatible ceramic materials based hydroxyapatite and metallic biomaterials 316L. *Appl. Surf. Sci.* **2013**, *285*, 65–71. [CrossRef]
6. Kannan, S.; Balamurugan, A.; Rajeswari, S. Hydroxyapatite coatings on sulfuric acid treated type 316L SS and its electrochemical behaviour in Ringer's solution. *Mater. Lett.* **2003**, *57*, 2382–2389. [CrossRef]
7. Kuroda, D.; Hiromoto, S.; Hanawa, T.; Katada, Y. Corrosion behavior of nickel-free high nitrogen austenitic stainless steel in simulated biological environments. *Mater. Trans.* **2002**, *43*, 3100–3104. [CrossRef]
8. Wang, L.; Zhao, X.; Ding, M.H.; Zheng, H.; Zhang, H.S.; Zhang, B.; Li, X.Q.; Wu, G.Y. Surface modification of biomedical AISI 316L stainless steel with zirconium carbonitride coatings. *Appl. Surf. Sci.* **2015**, *340*, 113–119. [CrossRef]
9. Schneider, J.M.; Rebholz, C.; Voevodin, A.A.; Leyland, A.; Matthews, A. Deposition and characterization of nitrogen containing stainless steel coatings prepared by reactive magnetron sputtering. *Vacuum* **1996**, *47*, 1077–1080. [CrossRef]
10. Kappaganthu, S.R.; Sun, Y. Formation of an MN-type cubic nitride phase in reactively sputtered stainless steel-nitrogen films. *J. Cryst. Growth* **2004**, *267*, 385–393. [CrossRef]
11. Kappaganthu, S.R.; Sun, Y. Influence of sputter deposition conditions on phase evolution in nitrogen-doped stainless steel films. *Surf. Coat. Technol.* **2005**, *198*, 59–63. [CrossRef]
12. Campoccia, D.; Montanaro, L.; Arciola, C.R. A review of the biomaterials technologies for infection-resistant surfaces. *Biomaterials* **2013**, *34*, 8533–8554. [CrossRef] [PubMed]
13. Ávalos, A.; Haza, A.I.; Morales, P. Nanopartículas de plata: Aplicaciones y riesgos tóxicos para la salud humana y el medio ambiente. *Rev. Complut. Cienc. Vet.* **2013**, *7*, 1–23. [CrossRef]
14. Taglietti, A.; Arciola, C.R.; D'Agostino, A.; Dacarro, G.; Montanaro, L.; Campoccia, D.; Cucca, L.; Vercellino, M.; Poggi, A.; Pallavicini, P.; et al. Antibiofilm activity of a monolayer of silver nanoparticles anchored to an amino-silanized glass surface. *Biomaterials* **2014**, *35*, 1779–1788. [CrossRef] [PubMed]
15. Velasco, S.C.; Cavaleiro, A.; Carvalho, S. Functional properties of ceramic-Ag nanocomposite coatings produced by magnetron sputtering. *Prog. Mater. Sci.* **2016**, *84*, 158–191. [CrossRef]
16. Manninen, N.K.; Ribeiro, F.; Escudeiro, A.; Polcar, T.; Carvalho, S.; Cavaleiro, A. Influence of Ag content on mechanical and tribological behavior of DLC coatings. *Surf. Coat. Technol.* **2013**, *232*, 440–446. [CrossRef]
17. Dang, C.; Li, J.; Wang, Y.; Yang, Y.; Wang, Y.; Chen, J. Influence of Ag contents on structure and tribological properties of TiSiN-Ag nanocomposite coatings on Ti–6Al–4V. *Appl. Surf. Sci.* **2017**, *394*, 613–624. [CrossRef]
18. Macías, H.A.; Yate, L.; Coy, L.E.; Olaya, J.; Aperador, W. Effect of nitrogen flow ratio on microstructure, mechanical and tribological properties of $TiWSiN_x$ thin film deposited by magnetron co-sputtering. *Appl. Surf. Sci.* **2018**, *456*, 445–456. [CrossRef]
19. *ASTM G99 Standard Test Method for Wear Testing with a Pin-on-Disk Apparatus*; ASTM International: West Conshohocken, PA, USA, 2006.
20. Ju, H.; He, S.; Yu, L.; Asempah, I.; Xu, J. The improvement of oxidation resistance, mechanical and tribological properties of W_2N films by doping silicon. *Surf. Coat. Technol.* **2017**, *317*, 158–165. [CrossRef]

21. Sánchez-López, J.C.; Abad, M.D.; Carvalho, I.; Galindo, R.E.; Benito, N.; Ribeiro, S.; Henriquese, M.; Cavaleirof, A.; Carvalho, S. Influence of silver content on the tribomechanical behavior on Ag-TiCN bioactive coatings. *Surf. Coat. Technol.* **2012**, *206*, 2192–2198. [CrossRef]
22. Bai, L.; Hang, R.; Gao, A.; Zhang, X.; Huang, X.; Wang, Y.; Tang, B.; Zhao, L.; Chu, P.K. Nanostructured titanium–silver coatings with good antibacterial activity and cytocompatibility fabricated by one-step magnetron sputtering. *Appl. Surf. Sci.* **2015**, *355*, 32–44. [CrossRef]
23. Aouadi, S.M.; Debessai, M.; Filip, P. Zirconium nitride/silver nanocomposite structures for biomedical applications. *J. Vac. Sci. Technol. B* **2004**, *22*, 1134–1140. [CrossRef]
24. Ferreri, I.; Lopes, V.; Tavares, C.J.; Cavaleiro, A.; Carvalho, S. Study of the effect of the silver content on the structural and mechanical behavior of Ag–ZrCN coatings for orthopedic prostheses. *Mater. Sci. Eng. C* **2014**, *42*, 782–790. [CrossRef] [PubMed]
25. Dahm, K.L.; Dearnley, P.A. On the nature, properties and wear response of s-phase (nitrogen-alloyed stainless steel) coatings on AISI 316L. *Proc. Inst. Mech. Eng. Part L J. Mater. Des. Appl.* **2000**, *214*, 181–198.
26. Bourjot, A.; Foos, M.; Frantz, C. Basic properties of sputtered 310 stainless steel—Nitrogen coatings. In *Metallurgical Coatings and Thin Films 1990*; Sartwell, B.D., Zemel, J.N., McGuire, G.E., Bresnock, F.N., Eds.; Elsevier: London, UK, 1990; pp. 533–542.
27. Thornton, J.A. The microstructure of sputter-deposited coatings. *J. Vac. Sci. Technol. A* **1986**, *4*, 3059–3065. [CrossRef]
28. Kelly, P.J.; Li, H.; Benson, P.S.; Whitehead, K.A.; Verran, J.; Arnell, R.D.; Iordanova, I. Comparison of the tribological and antimicrobial properties of CrN/Ag, ZrN/Ag, TiN/Ag, and TiN/Cu nanocomposite coatings. *Surf. Coat. Technol.* **2010**, *205*, 1606–1610. [CrossRef]
29. Leyland, A.; Matthews, A. On the significance of the H/E ratio in wear control: A nanocomposite coating approach to optimised tribological behaviour. *Wear* **2000**, *246*, 1–11. [CrossRef]
30. Buršík, J.; Buršíková, V.; Souček, P.; Zábranský, L.; Vašina, P. Characterization of Ta-BC nanostructured hard coatings. *IOP Conf. Ser. Mater. Sci. Eng.* **2017**, *175*, 012020. [CrossRef]
31. Pöhl, F.; Hardes, C.; Theisen, W. Scratch behavior of soft metallic materials. *AIMS Mater. Sci.* **2016**, *3*, 390–403. [CrossRef]
32. Tschiptschin, A.P.; Garzon, C.M.; Lopez, D.M. Scratch resistance of high nitrogen austenitic stainless steels. In *Tribology and Interface Engineering Series*; Sinha, S., Ed.; Elsevier: Amsterdam, The Netherlands, 2006; Volume 51, pp. 280–293.
33. *ASTM C1624-05 Standard Test Method for Adhesion Strength and Mechanical Failure Modes of Ceramic Coatings by Quantitative Single Point Scratch Testing*; ASTM International: West Conshohocken, PA, USA, 2005.
34. Mo, J.L.; Zhu, M.H. Tribological oxidation behaviour of PVD hard coatings. *Tribol. Int.* **2009**, *42*, 1758–1764. [CrossRef]
35. Yu, X.; Qin, Y.; Wang, C.B.; Yang, Y.Q.; Ma, X.C. Effects of nanocrystalline silver incorporation on sliding tribological properties of Ag-containing diamond-like carbon films in multi-ion beam assisted deposition. *Vacuum* **2013**, *89*, 82–85. [CrossRef]
36. Ju, H.; Yu, D.; Yu, L.; Ding, N.; Xu, J.; Zhang, X.; Zheng, Y.; Yang, L.; He, X. The influence of Ag contents on the microstructure, mechanical and tribological properties of ZrN-Ag films. *Vacuum* **2018**, *148*, 54–61. [CrossRef]

Article

Influence of Cu Content on the Structure, Mechanical, Friction and Wear Properties of VCN–Cu Films

Fanjing Wu [1,2], Lihua Yu [1,*], Hongbo Ju [1], Junhua Xu [1] and Ji Shi [2]

1 School of Materials Science and Engineering, Jiangsu University of Science and Technology, Zhenjiang 212003, China; fanjingwu6@163.com (F.W.); hbju@rocketmail.com (H.J.); jhxu@just.edu.cn (J.X.)

2 Department of Science and Technology, Tokyo Institute of Technology, Tokyo 152-8552, Japan; shi.j.aa@m.titech.ac.jp

* Correspondence: lhyu6@just.edu.cn; Tel.: +86-511-8441-1035; Fax: +86-511-8440-1184

Received: 21 January 2018; Accepted: 3 March 2018; Published: 7 March 2018

Abstract: VCN–Cu films with different Cu contents were deposited by reactive magnetron sputtering technique. The films were evaluated in terms of their microstructure, elemental composition, mechanical, and tribological properties by X-ray diffraction (XRD), energy-dispersive X-ray spectroscopy (EDS), high resolution transmission electron microscopy (HR-TEM), Raman spectrometry, nano-indentation, field emission scanning electron microscope (FE-SEM), Bruker three-dimensional (3D) profiler, and high-temperature ball on disc tribo-meter. The results showed that face-centered cubic (fcc) VCN, hexagonal close-packed (hcp) V_2CN, fcc-Cu, amorphous graphite and CN_x phases co-existed in VCN–Cu films. After doping with 0.6 at.% Cu, the hardness reached a maximum value of ~32 GPa. At room temperature (RT), the friction coefficient and wear rate increased with increasing Cu content. In the temperature range of 100–500 °C, the friction coefficient decreased, but the wear rate increased with the increase of Cu content.

Keywords: VCN–Cu films; microstructure; mechanical; friction property; wear property

1. Introduction

Transition metal nitride (TMN) nanocomposite films, exhibiting a good combination of mechanical, chemical, and wear resistance properties [1–3], have been widely used in many fields (molds, cutting tools, engine parts, etc.) for surface strengthening [4,5]. Vanadium nitride (VN) films are a good alternative for their more common TiN, ZrN, and CrN counterparts [6–9], due to enhanced selflubricating ability [10]. In our previous study [11], a series of VCN films with different carbon content were deposited onto a Si substrate using the reactive magnetron sputtering method. The mechanical and room temperature tribological properties of VN-based films can be improved greatly by doping with carbon. For example, the hardness of VCN films reached 33 GPa [12], being higher than that of binary VN (~20 GPa) and VC (20–22 GPa) films [13], the room-temperature tribological properties reached the lowest value of 0.22 [12], being lower than VN (~0.47) and VC (~0.35) films [13]. However, at medium temperature, the friction coefficients were still relatively high and this would limit its further application.

Recently, TMN nanocomposite films containing soft metals (such as copper and silver) have been a hot topic due to their potential applications in different fields, such as wear, friction, corrosion protection, medicine, optics, and so on. Much research focuses on the nitrides, carbides, and oxides of transition metals in copper doping, mainly studied their tribological properties [1,5,14]. Besides, some researchers also have conducted a thorough study on wear [15,16]. Although at room temperature, the available results regarding the effect of soft metal Cu on the friction properties and wear of thin films are rather contradictory [1,5,12]. The incorporation of Cu was shown to be an effective approach to reduce friction, especially for elevated temperature. For instance, Suszko et al. [5] deposited

Mo_2N–Cu nanocomposite films to study the effect of copper on the tribological properties in the range of 25–400 °C. They found that, after adding some Cu, the friction coefficient of Mo_2N film was reduced obviously in the temperature range of 200–400 °C. Yu et al. [17] prepared W_2N–Cu films and studied the effect of Cu content on the tribological properties in the temperature range of 25–600 °C. They found that the incorporation of Cu could improve the friction coefficient of W_2N film in this temperature range. Therefore, the incorporation of soft phased Cu into VCN films is an effective method to improve the medium-temperature tribological properties.

To the best of our knowledge, VCN–Cu coatings were not investigated to date. Therefore, the present study aimed at preparing, characterizing, and investigating the tribological properties from 25 to 500 °C of VCN–Cu films. To understand processes that are involved into friction and wear of VCN–Cu films, the effect of Cu on the microstructure, Micro-hardness and residual stress of the prepared VCN films was thoroughly studied.

2. Experimental Details

2.1. Preparation of Films

VCN–Cu composite films with a thickness of about 2 μm were deposited on 304 stainless steels and Si (100) wafers using a multi-target magnetron sputtering system (JGP-450, SKY, Shenyang, China). Water-cooled 75-mm-diameter V (99.9%), C (graphite) (99.9%), and Cu (99.99%) targets were corresponding installed in three cathodes. Si (100) wafers were used as substrate materials for structural and mechanical investigations. Stainless steel 304 (15 mm × 15 mm × 2 mm) were used as substrate materials for dry sliding ball-on-disk tests. The 304 stainless steel were polished with sandpaper and soft cloth used to clean the surface to removed residual scratches. Before being placed on the substrate holder in the chamber, the substrates were cleaned with successive rinses in ultrasonic baths of acetone and alcohol, and blown dry with dry air. The base pressure was 6.0×10^{-4} Pa before deposition. Just prior to initiating deposition, 10 min sputter clean of the targets and 15 min deposition of V (target power, 200 W) were conducted in a pure Argon atmosphere as the transition layer to confirm a good adhesion force of the films. Then, VCN–Cu films with various Cu content were achieved by fixing the powers of V target at 200 W and C (graphite) target at 90 W and adjusting the power of Cu target from 0 to 25 W. The target-substrate distance was 78 mm with no substrate bias voltage, no substrate temperature and a constant working pressure (0.3 Pa) with the same ratio of Ar and N_2 flow rates (10:7) in a radio-frequency magnetron sputtering system. Detailed preparation parameters in Table 1.

Table 1. The preparation parameters of VCN–Cu composite films.

VCN–Cu Composite Films	Experimental Parameters
Base pressure	<6.0×10^{-4} Pa
Total pressure	0.3 Pa
Ar flow rate	10 sccm
N_2 flow rate	7 sccm
Base rotation speed	3 r/min
V target power	200 W
C target power	90 W
Cu target power	0 W, 10 W, 15 W, 20 W, 25 W
Negative bias of the substrate	–
Deposition temperature	Room temperature
Deposition time	2 h

2.2. Characterization of the Films

The phases of VCN–Cu films were explored by X-ray diffraction (XRD, Shimazu-6000, Shimadzu, Kyoto, Japan) with a Cu Kα source, operated at 40 kV and 35 mA with a scanning speed was

4°/min. The average grain crystal size (D_c) of VCN–Cu was calculated by Debye-Scherer, as seen in equation [18]:

$$D_c = \frac{0.89\lambda}{B\cos\theta} \quad (1)$$

where λ is X-ray wavelength, B is full width at half maximum (FWHM) of diffraction peak and θ represents the diffraction angle. The diffraction profiles were corrected for the instrumental broadening using an alumina sample, which had large crystallites and were free from the defects. The cross-sections of the films were observed by a field emission scanning electron microscopy (FE-SEM, Merlin Compact-6170, ZEISS, Jena, Germany). A high-resolution transmission electron microscopy (HR-TEM, JEOL JEM-2010F, JEOL, Tokyo, Japan) that was operated at an accelerating voltage of 200 KV and energy dispersive spectroscopy (EDS) on an EDAX-DX-4 energy dispersive analyzer were used to characterize the crystallinity, microstructure, and elemental composition of the film. A micro-Raman spectrometer (inVia*, Renishaw, Gloucestershire, UK) having a 514.5 nm Ar^+ laser as an excitation source with back scattering geometry was used in recording Raman spectra at room temperature.

Hardness of the films were determined by nano-indenter CPX + NHT_2 + MST (CSM, Peseux, Switzerland) equipped with a diamond Berkovich indenter tip (3-side pyramid). In order to minimize the substrate's influence on the hardness of the films, a maximum load of 3 mN was used to meet the $d/h < 0.1$ criterion (where d and h are the indenter penetration depth into the film and the film thickness, respectively) using an automatic indentation mode that was programmed to place indentations in a 3×3 array. Nine points were selected for each sample and averaged. An automatic indentation mode programmed to place indentations in a 3×3 array. Each sample was tested three times. Fused silica was used as a reference sample for the calibration before indents were made.

The tribological properties were conducted along a circular track (diameter: 8 mm) against Al_2O_3 counterpart (diameter: 9.38 mm) for 30 min. A constant speed of 50 rpm was maintained with a normal load of 3 N in an atmosphere with a relative humidity of about 25%–30%. The temperature was fixed at 25, 100, 200, 300, 400, and 500 °C, respectively. Friction tests were performed three times under the same conditions for the same batch of samples, and the accuracy of the experiment was confirmed based on the repeatability of the experimental results. Thereafter, a profilometer (Bruker DEKTAK-XT, Billerica, MA, USA) was used to determine wear volume loss of the films (V) through examining the worn tracks. In order to calculate the wear rates (W) of the as-deposited films, an Archard's classical wear equation [19] was listed as follows:

$$W - \frac{V}{S \times L} \quad (2)$$

here S represents the sliding distance, L is the loading force, and V is wear volume. After wear tests, the wear tracks were also detected by XRD (Shimazu-6000, Shimadzu, Kyoto, Japan). A spot size of 1.0 mm was fixed on the wear track, which was less than the wear track width, hence our ability to probe the wear track with the XRD.

The average stress σ in the films was determined by means of Stoney's formula [20] (Equation 3).

$$\sigma = \frac{E_s t_s^2\left(\frac{1}{r} - \frac{1}{R}\right)}{6(1 - V_s)t_f} \quad (3)$$

where the subscript S refers to the substrate, E_S and V_S are Young's modulus and Poisson's ratio of silicon substrate, respectively, t_f and t_s are thicknesses of film and substrate, respectively, R and r are the substrate curvature radii of the original Si wafer and the coated Si wafer, which was determined by measuring the curvature of the silicon wafer before and after deposition by Bruker three-dimensional (3D) Profiler (Bruker DEKTAK-XT, Bruker, Billerica, MA, USA).

3. Results and Discussion

3.1. Composition and Phase Structure

The chemical composition of the VCN–Cu composite films with varying Cu target powers are summarized in Table 2. As listed, an increase in Cu target power from 0 to 25 W have a corresponding increase in Cu, while the relative content of V and C decreases but the content of N remains relatively uniform (~50 at.%.) irrespective of the Cu target power. The accuracy of the carbon and nitrogen content is limited by the EDS method.

Table 2. The composition of VCN–Cu films.

Power of Copper Target (W)	Chemical Composition (at.%)			
	V	C	N	Cu
0	31.5 ± 2.2	16.9 ± 0.7	51.6 ± 2.8	0
10	31.2 ± 2.0	16.7 ± 0.7	51.5 ± 2.7	0.6 ± 0.1
15	30.6 ± 1.9	16.6 ± 0.6	51.1 ± 2.5	1.7 ± 0.2
20	26.9 ± 1.8	13.9 ± 0.5	50.7 ± 2.5	8.5 ± 0.3
25	24.3 ± 1.6	12.6 ± 0.4	49.9 ± 2.4	13.2 ± 0.4

XRD patterns of VCN–Cu films with different Cu contents are presented in Figure 1. As shown, XRD patterns of VCN–Cu films with different Cu contents are presented in Figure 1. As shown, for VCN film, three peaks are detected, which are corresponding to the face-centered cubic (fcc) VN (111) (JCPDF 35-0768), hexagonal close-packed (hcp) V_2N (111) (JCPDF 32-1413) and fcc-VN (220) (JCPDF 35-0768), where some N atoms in VN and V_2N lattices are replaced by C atoms. The incorporation of Cu does not change the crystal structure. The preferred orientation of the films is (111). Moreover, no diffraction peaks corresponding to Cu and Cu compounds are detected. We utilized other characterization techniques to analyze the Cu phase in this paper later. A contribution to the shift of peaks, maybe due to the compressive stress in the VCN–Cu films [21].

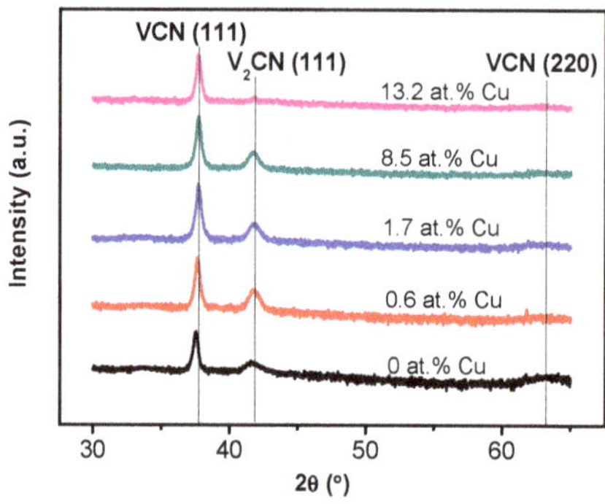

Figure 1. X-ray diffraction (XRD) patterns of VCN–Cu films with different Cu contents.

In order to obtain further information about the carbon containing phase, the Raman spectrum test of VCN–Cu films are conducted, and the results are shown in Figure 2. As shown, a resolved peak located at 1560 cm^{-1} can be ascribed to the G (graphitic) absorption band of carbon [22]. Two peaks are detected at 1506 cm^{-1} and 1720 cm^{-1}, which are associated with C–N and C=N absorption bands of CN_x [23]. The formation of amorphous will produce higher interface energy. When compared with amorphous graphite, VCN solid solution are more easily formed under the same conditions due to the minimum energy principle. After the solid solution are saturated, amorphous graphite phases begin to form. As shown in Table 2, the V:C ratio is always about 2:1, and the G (graphitic) absorption peak detected in Figure 2 represents that the VCN solid solution have already been saturated at 0 at.% Cu. Therefore, as the Cu content increases from 0 at.% to 13.21 at.%, the Raman peaks do not change significantly.

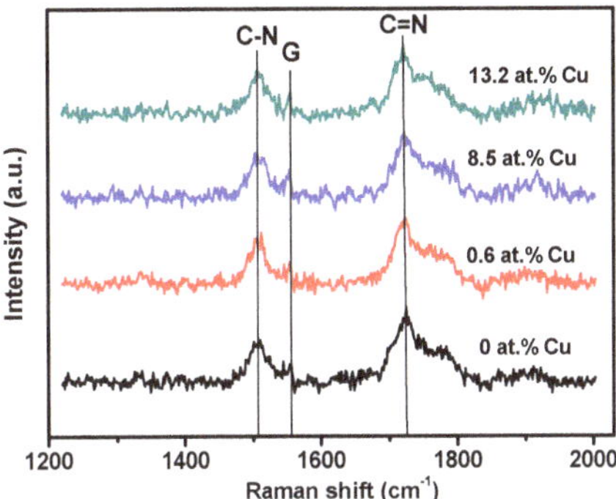

Figure 2. Raman Patterns of VCN–Cu composited films with different Cu contents.

In order to further confirm the state of Cu in the VCN–Cu films, TEM, HR-TEM images and corresponding selected area electron diffraction (SAED) patterns of VCN–1.7 at.% Cu and VCN–13.2 at.% Cu films are shown in Figures 3 and 4.

As shown in Figure 3a,b, three sets of lattice fringes with lattice spacing of about 0.240, 0.218 and 0.208 nm (standard value: $d_{(\text{fcc-VN (111)})}$ = 0.238 nm; $d_{(\text{fcc-V}_2\text{N (111)})}$ = 0.215 nm; $d_{(\text{fcc-Cu (111)})}$ = 0.208) are observed, which is corresponding to fcc-VCN (111), hcp-V_2CN (111), and fcc-Cu (111). Figure 3c is selected area electron diffraction (SAED) pattern of VCN–1.7 at.% Cu film, showing fcc-VCN (111), hcp-V_2CN (111), fcc-Cu (111), and fcc-VCN (220) diffraction rings.

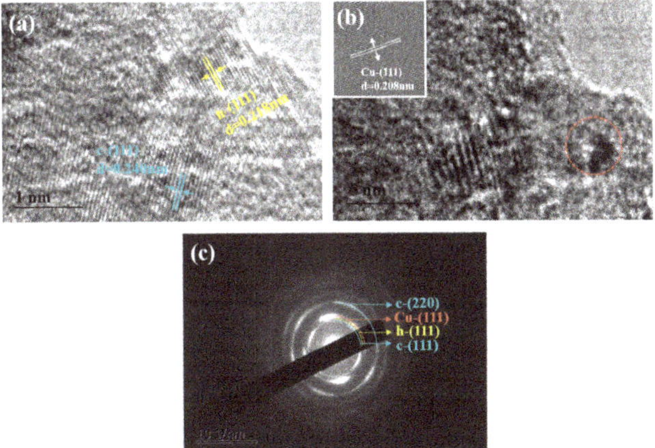

Figure 3. (**a**) Selected area electron diffraction (SAED) pattern and (**b**,**c**) high resolution transmission electron microscopy (HR-TEM) observations of VCN–1.7 at.% Cu film.

As shown in Figure 4a, the VCN–Cu film exhibits a columnar crystal texture, also showing that the adding Cu particles are well dispersed in the film. Fast Fourier Transform (FFT) are carried out on the film in Figure 4b. The lattice fringes with lattice spacing of 0.208 and 0.240 nm are detected, these value are almost the same as the calculated HR-TEM value for fcc-Cu (111) and fcc-VCN (111). Ezirmik et al. [24] prepared CrN–Cu composite films found that crystalline Cu distributed in the grain boundary of the CrN film.The corresponding SAED pattern in Figure 4c also show the diffraction rings of fcc-VCN, hcp-V_2CN, and fcc-Cu. Thus, Cu probably distributed in the VCN grain boundaries in this study.

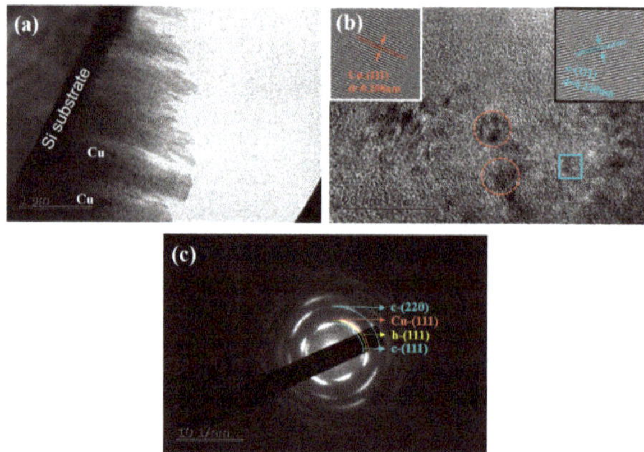

Figure 4. (**a**) TEM, (**b**) SAED pattern, and (**c**) HR-TEM observations of VCN–13.2 at.% Cu film.

Based on the classical Debye-Scherer equation, the average grain sizes were calculated according to the values of λ and *B* listed in Table 3. Three measurements were made for each value. The results are shown in Figure 5, the grain size of VCN–Cu decreases gradually with the increase of Cu content, and reached a minimum value of 11.65 nm at 13.2 at.% Cu. This is due to VCN grains are forced to re-nucleate on the top of Cu and the presence of Cu in the grain boundaries could also inhibit the VCN grains growth [5,25]. In Figure 1, the Cu phase was not detected even at the highest Cu content of 13.2 at.% , which is due to the large number of small sizes of Cu grains dispersed in the film. For XRD, it is not enough to meet the crystallographic plane of the Bragg condition. But, in the Figure 3a, the diffraction ring of Cu can be detected even with 1.7 at.% Cu.

Table 3. The values of λ and *B* obtained from XRD patterns.

Cu Content	λ (nm)	$B_{VN\ (111)}$ (°)
0 at.%		0.400 ± 0.002
0.6 at.%		0.448 ± 0.003
1.7 at.%	0.154	0.534 ± 0.002
8.5 at.%		0.619 ± 0.001
13.2 at.%		0.713 ± 0.002

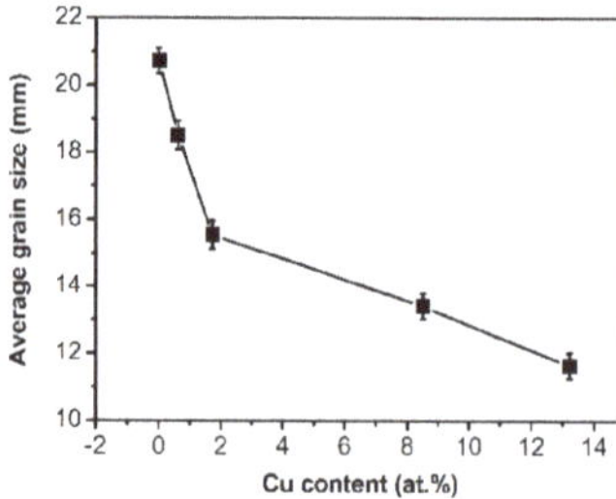

Figure 5. Grain sizes of VCN–Cu composite films with different Cu contents.

Based on the analysis of XRD, Raman, and HR-TEM, the VCN–Cu films consist of fcc-VCN, hcp-V_2CN, fcc-Cu phase, and amorphous graphite and CN_x phase. Besides, the Cu particles are disperse in the films.

3.2. Mechanical Properties

The hardness (H) and residual stress (σ) of VCN–Cu films with different Cu contents are shown in Figure 6. The H of ternary VCN film is 28.0 GPa. The hardness increases slightly then decreases with a maximum value of 32 GPa at 0.6 at.% Cu. Meanwhile, the residual stress increases then decreases with a maximum value of −1.94 GPa at 0.6 at.% Cu (the negative sign indicates that the residual stress is compressive).

According to the analysis of microstructure, VCN (H_{VN}: ~21 GPa [26]), V_2CN (H_{V_2N}: 25 GPa [26]), Cu (3 GPa [5]), graphite C (2 GPa [4]), and CN_x (15.8 GPa [27]) consist in VCN–Cu films. With the increase of Cu content, the Cu phase increases while other phases decrease per unit area.

There are several models that can probably affect the hardness: solid solution strengthening, "Hall-Patch" relationship (refined crystalline strengthening), stress intensification, dispersion strengthened, and so on. As the content of Cu was 0 at.%, the hardness of VCN in this study are higher than that of VN. The reason is that the solid solution leads to lattice distortion which formed the alternating stress field, thus bring a higher hardness. Since the C has reached saturation at 0 at.% Cu, the effect of solid solution strengthening on all types of VCN–Cu films is the same. As the content of Cu reaches 0.6 at.%, the hardness enhancement was mainly attributed to the effect of fine grain strengthening and residual stress [5,27]. At this point, the compressive stress reaches the maximum value of −1.94 GPa. As the content of Cu continue increase to 13.2 at.%, the decrease of hardness is mainly due to a drop of VCN and V_2CN phases and compressive stress value, and a rise of low hardness soft Cu phase. All of these factors simultaneously resulted in a decline in the hardness of films. However, Cu particles are well disperse in the VCN matrix, but the movement of dislocations cannot be impeded due to softer Cu phase, thus the dispersion strengthened mechanism fail to apply in our paper [17,28].

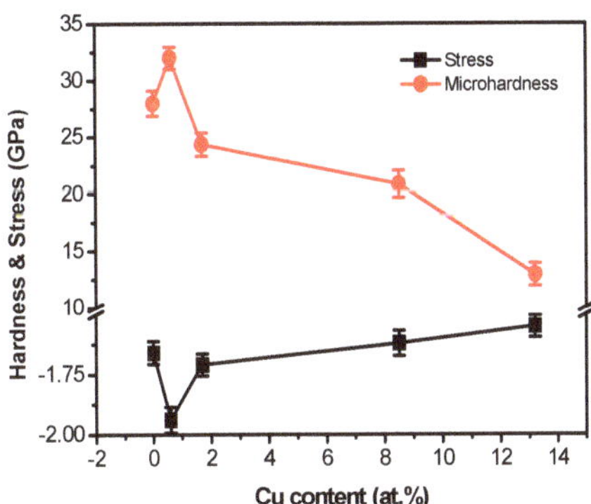

Figure 6. Micro-hardness and stress of VCN–Cu composite films with different Cu contents.

3.3. Friction and Wear Properties

3.3.1. Friction and Wear Properties at Room Temperature

Figure 7 shows the average friction coefficients and wear rate of VCN–Cu films with different Cu contents at room temperature. As shown, the average friction coefficient and wear rate of VCN film are ~0.42 and 0.193 × 10^{-7} $mm^3 \cdot N^{-1} \cdot m^{-1}$. With the increase of Cu content, the average friction coefficient, and wear rate increase gradually. As the content of Cu is 13.2 at.%, the friction coefficient and wear rate reach the maximum values of 0.63 and 1.21 × 10^{-7} $mm^3 \cdot N^{-1} \cdot m^{-1}$, respectively.

At room temperature, the relatively high CoF of the Cu-doped VCN films can be explained by the decrease of V and C content, which provide the lubrication phase. The wear rate of the films are mainly affected by its mechanical properties [5,28]. The decrease of the hardness of the film would lead to the increase of the contact area between the film and the friction pair. This subsequently causes the deformation of the film and shearing to form wear debris. Then wear rate increase [29,30]. Besides, the addition of copper weakens the bonding strength between grains. Zhang et al. [31] had reported that the surface of the film tended to produce physical adsorption and leaded to the falling off of the other nanoparticles with the increase of Cu. The falling spherical particles roll on the friction surface to cause abrasive wear and increased the wear rate. The more Cu added, the easier it is to form abrasive wear.

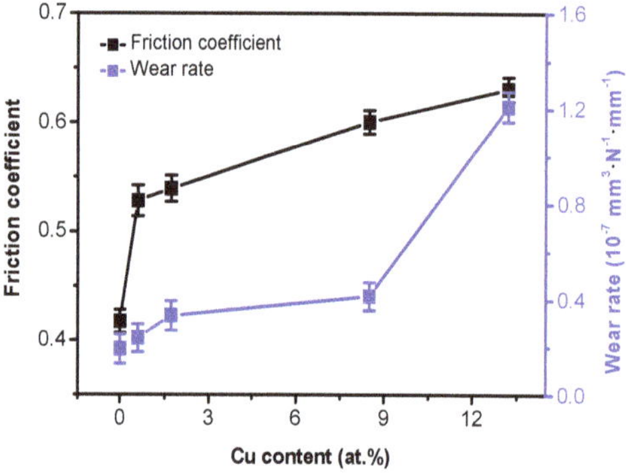

Figure 7. Average coefficient curve and wear rate of VCN–Cu films with different Cu contents at RT.

3.3.2. Friction and Wear Properties at Elevated Temperature

The friction coefficients and wear rates of VCN, VCN–1.7 at.% Cu and VCN–13.2 at.% Cu films in the temperature range of 100–500 °C are presented in Figure 8. As shown, as the temperature is constant, the friction coefficient decreases, but the wear rate increases with the increase of Cu content. As Cu content is constant, the friction coefficient first increases and then decreases while the wear rate increases with the increase of the temperature.

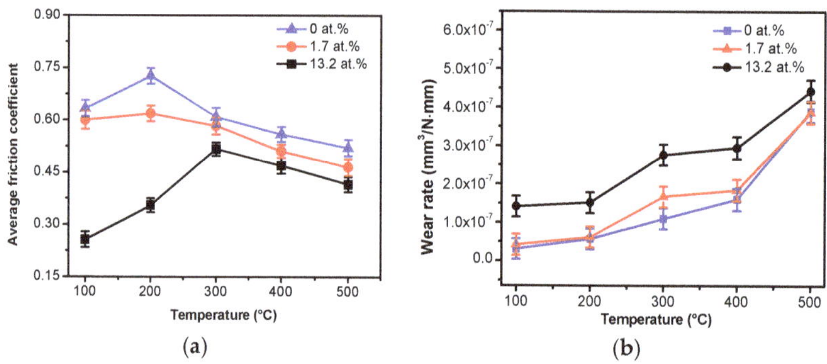

Figure 8. (**a**) Friction coefficient and (**b**) wear rate of VCN–Cu films at different temperatures.

The XRD patterns of the wear track after sliding at different temperatures are shown in Figure 9. As shown in Figure 9a, Cu and CuO are detected in all the test samples, but VO_2 and Cu_xO are only detected at 400 °C. At 500 °C, V_2O_5 is also detected. As shown in Figure 9b, at 500 °C, V_2O_5 and VO_2 are detected for VCN film. Besides, Cu_xO and Cu are also detected for VCN–Cu films. In addition, with the increase of Cu content, the peak intensities of the V_2O_5 and VO_2 decrease, the peak intensities of Cu_xO and Cu increase. This suggests that the content of V_2O_5 and VO_2 phases on the wear track decrease and the content of Cu_xO and Cu increase.

In order to study the effect of oxides on the tribological properties at different temperatures, the relative content of phases existed in the films are calculated by using "Adiabatic method" [17]. The results are shown in Figure 10. As shown in Figures 9a and 10a, the content of (VCN + V_2CN) decreased, while the content of Cu, CuO, VO_2, Cu_xO, and V_2O_5 increase gradually with the increase of the temperature. As shown in Figures 9b and 10b, the content of CuO and Cu_xO obviously increase, while the content of VO_2 and V_2O_5 decrease gradually.

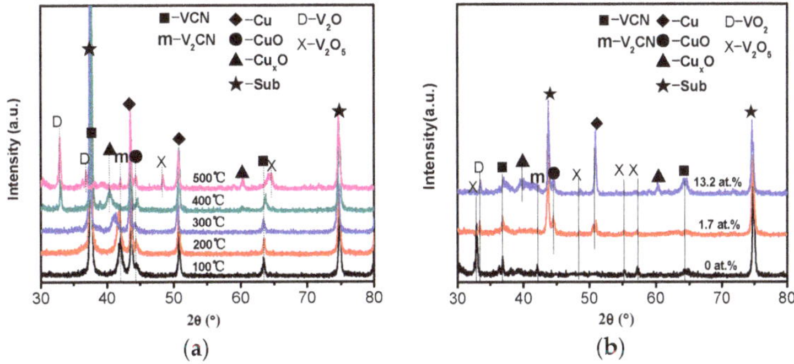

Figure 9. XRD patterns of VCN–13.2 at.% Cu composite films after sliding (**a**) at different temperatures and (**b**) VCN–Cu composite films after sliding at 500 °C.

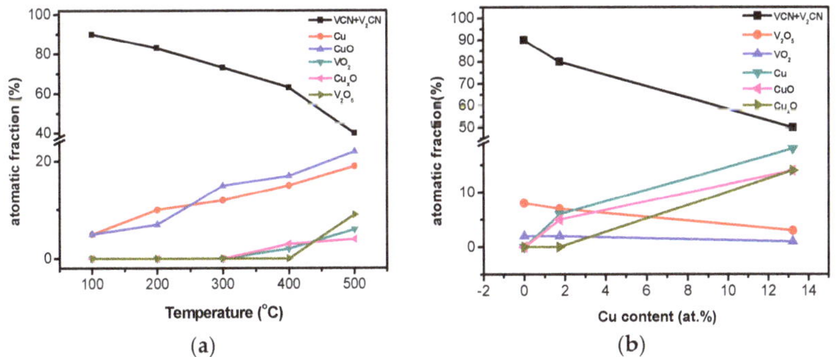

Figure 10. Oxide-content of VCN–13.2 at.% Cu composite films after sliding (**a**) at different temperatures and (**b**) oxide-content of VCN–Cu composite films after sliding at 500 °C.

As the testing temperature increased, the water vapor in the environment disappeared, and C and CN_x phase generated CO_2 [32]. The increase in friction coefficient between 25 and 300 °C was mainly due to the sp^2 structure of C been gradually destroyed leading to the disappearance of their role as solid lubricant, combined with the disappearance of the adsorption on the films surface [17,32]. Meanwhile the content of Magneli phase V_2O_5 is relatively low, hence, cannot effectively reduce the friction. Between 300 and 500 °C, the friction coefficient of VCN–Cu composite films decrease significantly. The content of Magneli phases V_2O_5 and VO_2 [33] increase enough to reduce the friction coefficient. In addition, Erdemir et al. [34] pointed out that when the oxides were more than one in the high-temperature friction system, the larger difference in the ionization potential of the oxides, the better lubrication effect. The ionization potentials of CuO and V_2O_5 are 2.74 [29] and 8.4 [34]. The difference between CuO and V_2O_5 could decrease the friction coefficient of the films. The increase of wear rate is mainly due to the metal adhesive that was caused by copper on the surface [17,31], the increase of oxides (CuO_x and VO_x), which can be worn away easily by the counterpart during the wear test due to its low shear strength.

The SEM images of the wear track of VCN and VCN–13.2 at.% Cu films at 300 and 500 °C are presented in Figure 11. When compared with 300 °C, the wear scars of the films become wider and deeper after sliding at 500 °C. At high temperatures, oxidation of the film is intensified and some wear debris (particles) are found in the wear track. When compared with VCN, more furrows and deeper wear marks are observed for VCN–Cu film at the same temperature.

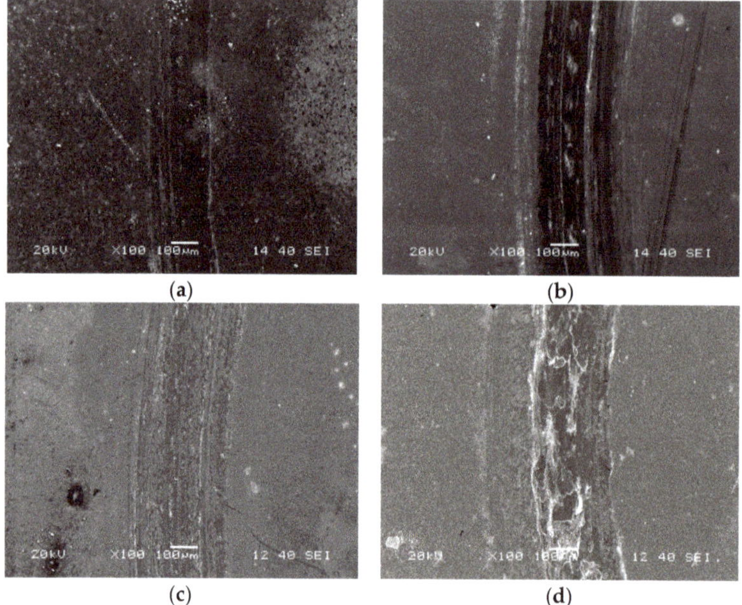

Figure 11. SEM images of the wear tracks of VCN–13.2 at.% Cu and VCN films at different temperatures: (**a**) VCN–Cu, 300 °C; (**b**) VCN–Cu, 500 °C; (**c**) VCN, 300 °C; (**d**) VCN, 500 °C.

4. Conclusions

In summary, VCN–Cu film was deposited using reactive magnetron sputtering:

- VCN–Cu films were consisted of a mixture of fcc-VCN, hcp-V_2CN, fcc-Cu, amorphous graphite and CN_x phase.
- The hardness of VCN–Cu films first increased and then decreased with the increase of Cu content.
- At room temperature, the friction coefficient and wear rate increased with an increasing Cu content.
- In the temperature range of 100–500 °C, the friction coefficient of VCN–Cu film was lower than VCN film, while the wear rate showed the opposite trend. The incorporation of Cu improved the friction properties, but failed to improve the wear resistance.

Acknowledgments: Especially thanks to the financial support from National Natural Science Foundation of China (Grant No. 51374115 and 51574131).

Author Contributions: Lihua Yu conceived and designed the experiments; Lihua Yu and Fanjing Wu performed the experiments and analyzed the data; Hongbo Ju and Junhua Xu contributed to data analysis; Lihua Yu wrote the paper; all authors participated and discussed this work and contributed to the submitted and published manuscript.

Conflicts of Interest: The authors declare no conflict of interest.

References

1. Pappacena, K.E.; Singh, D.; Ajayi, O.O.; Routbort, J.L.; Erilymaz, O.L.; Demas, N.G.; Chen, G. Residual stresses, interfacial adhesion and tribological properties of MoN/Cu composite coatings. *Wear* **2012**, *278*, 62–70. [CrossRef]
2. Stone, D.S.; Migas, J.; Martini, A.; Smith, T.; Muratore, C.; Voevodin, A.A.; Aouadi, S.M. Adaptive NbN/Ag coatings for high temperature tribological applications. *Surf. Coat. Technol.* **2012**, *206*, 4316–4321. [CrossRef]
3. Caicedo, J.C.; Zambrano, G.; Aperador, W.; Escobar–Alarcon, L.; Camps, E. Mechanical and electrochemical characterization of vanadium nitride (VN) thin films. *Appl. Surf. Sci.* **2011**, *258*, 312–320. [CrossRef]
4. Wu, F.; Yu, L.; Ju, H.; Asempah, I.; Xu, J. Structural, mechanical and tribological properties of NbCN–Ag nanocomposite films deposited by reactive magnetron sputtering. *Coatings* **2018**, *8*, 50. [CrossRef]
5. Suszko, T.; Gulbiński, W.; Jagielski, J. Mo_2N/Cu thin films–the structure, mechanical and tribological properties. *Surf. Coat. Technol.* **2006**, *200*, 6288–6292. [CrossRef]
6. Su, Y.; Yao, S.; Leu, Z.; Wei, C.; Wu, C. Comparison of tribological behavior of three films—TiN, TiCN and CrN—Grown by physical vapor deposition. *Wear* **1997**, *213*, 165–174. [CrossRef]
7. Polcar, T.; Kubart, T.; Novák, R.; Kopecký, L.; Široký, P. Comparison of tribological behaviour of TiN, TiCN and CrN at elevated temperatures. *Surf. Coat. Technol.* **2005**, *193*, 192–199. [CrossRef]
8. Auger, M.A.; Araiza, J.J.; Falcony, C.; Sánchez, O.; Albella, J.M. Hardness and tribology measurements on ZrN coatings deposited by reactive sputtering technique. *Vacuum* **2007**, *81*, 1462–1465. [CrossRef]
9. Santecchia, E.; Hamouda, A.M.S.; Musharavati, F.; Zalnezhad, E.; Cabibbo, M.; Spigarelli, S. Wear resistance investigation of titanium nitride–based coatings. *Ceram. Int.* **2015**, *41*, 10349–10379. [CrossRef]
10. Franz, R.; Mitterer, C. Vanadium containing self–adaptive low–friction hard coatings for high–temperature applications: A review. *Surf. Coat. Technol.* **2013**, *228*, 1–13. [CrossRef]
11. Yu, L.; Li, Y.; Ju, H.; Xu, J. Microstructure, mechanical and tribological properties of magnetron sputtered VCN films. *Surf. Eng.* **2017**, *33*, 919–924. [CrossRef]
12. Mu, Y.; Liu, M.; Zhao, Y. Carbon doping to improve the high temperature tribological properties of VN coating. *Tribol. Int.* **2016**, *97*, 327–336. [CrossRef]
13. Mitterer, C.; Fateh, N.; Munnik, F. Microstructure–property relations of reactively magnetron sputtered VC_xN_y films. *Surf. Coat. Technol.* **2011**, *205*, 3805–3809. [CrossRef]
14. Chao, M.; Rong, F. Formation and mechanism of the third body of copper under low and high friction speed. *Sci. Technol. Eng.* **2010**, *15*, 036. (In Chinese)
15. Maruda, R.W.; Feldshtein, E.E.; Legutko, S.; Królczyk, G.M. Improving the efficiency of running-in for a bronze–stainless steel friction pair. *J. Frict. Wear* **2015**, *36*, 548–553. [CrossRef]
16. Maruda, R.W.; Krolczyk, G.M.; Feldshtein, E.; Nieslony, P.; Tyliszczak, B.; Pusavec, F. Tool wear characterizations in finish turning of AISI 1045 carbon steel for MQCL conditions. *Wear* **2017**, *372*, 54–67. [CrossRef]
17. Yu, L.; Zhao, H.; Ju, H.; Xu, J. Influence of Cu content on the structure, mechanical and tribological properties of W_2N–Cu films. *Thin Solid Films* **2017**, *624*, 144–151. [CrossRef]
18. Xu, J.; Chen, J.; Yu, L. Influence of Si content on the microstructure and mechanical properties of VSiN films deposited by reactive magnetron sputtering. *Vacuum* **2016**, *131*, 51–57. [CrossRef]
19. Shi, J.; Muders, C.M.; Kumar, A.; Jiang, X.; Pei, Z.L.; Gong, J.; Sun, C. Study on nanocomposite Ti–Al–Si–Cu–N films with various Si contents deposited by cathodic vacuum arc ion plating. *Appl. Surf. Sci.* **2012**, *258*, 9642–9649. [CrossRef]
20. Ye, F.; Zhao, H.; Tian, X. Influence of niobium addition on structure and mechanical properties of W–Nb–N coatings. *Vacuum* **2017**, *144*, 8–13. [CrossRef]

21. Shtansky, D.V.; Bondarev, A.V.; Kiryukhantsev-Korneev, P.V.; Rojas, T.C.; Godinho, V.; Fernandez, A. Structure and tribological properties of MoCN–Ag coatings in the temperature range of 25–700° C. *Appl. Surf. Sci.* **2013**, *273*, 408–414. [CrossRef]
22. Bondarev, A.V.; Golizadeh, M.; Shvyndina, N.V.; Shchetinin, I.V.; Shtansky, D.V. Microstructure, mechanical, and tribological properties of Ag-free and Ag-doped VCN coatings. *Surf. Coat. Technol.* **2017**, *331*, 77–84. [CrossRef]
23. Carvalho, I.; Henriques, M.; Oliveira, J.C.; Alves, C.F.A.A.; Piedade, P.; Carvalho, S. Influence of surface features on the adhesion of Staphylococcus epidermidis to Ag–TiCN thin films. *Sci. Technol. Adv. Mater.* **2013**, *14*, 35009–35018. [CrossRef] [PubMed]
24. Ezirmik, V.; Senel, E.; Kazmanli, K.; Erdemir, A.; Urgen, M. Effect of copper addition on the temperature dependent reciprocating wear behaviour of CrN coatings. *Surf. Coat. Technol.* **2007**, *202*, 866–870. [CrossRef]
25. Tan, S.; Zhang, X.; Wu, X.; Fang, F.; Jiang, J. Comparison study on structure of Si and Cu doping CrN films by reactive sputtering. *Appl. Surf. Sci.* **2011**, *257*, 5595–5600. [CrossRef]
26. Sanjinés, R.; Hones, P.; Lévy, F. Hexagonal nitride coatings: electronic and mechanical properties of V_2N, Cr_2N and δ-MoN. *Thin Solid Films* **1998**, *332*, 225–229. [CrossRef]
27. Xu, X.; Zhuo, L.; Xia, D.; Wei, B.; Hao, X. Nanoindentation and friction/wear behaviors of magnetron sputtered CN_x/SiC bilayer films on Titanium (TA_2) substrate. *Mocaxue Xuebao* **2009**, *29*, 256–260. (In Chinese)
28. Xu, J.; Ju, H.; Yu, L. Effects of Mo content on the microstructure and friction and wear properties of TiMoN films. *Acta Metall. Sin.* **2012**, *48*, 1132–1138. (In Chinese) [CrossRef]
29. Yu, L.; Sun, C.; Xu, J. Cu Doping and improvement of mechanical and tribological properties of NbCN composite coating. *Chin. J. Vac. Sci. Technol.* **2016**, *36*, 377–384. (In Chinese)
30. Wei, Y.; Gong, C. Effects of pulsed bias duty ratio on microstructure and mechanical properties of TiN/TiAlN multilayer coatings. *Appl. Surf. Sci.* **2011**, *257*, 7881–7886. [CrossRef]
31. Zhang, Y.; Yan, J.; Sun, L.; Yang, G.; Zhang, Z.; Zhang, P. Friction reducing anti-wear and self-repairing properties of nano-Cu additive in lubricanting oil. *J. Mech. Eng.* **2010**, *46*, 74–79. (In Chinese) [CrossRef]
32. Xu, J.; Cao, J.; Yu, L. Microstructures, mechannical properties and friction properties of TiVCN composite films. *Acta Metall. Sin.* **2012**, *48*, 555–560. (In Chinese) [CrossRef]
33. Gassner, G.; Mayrhofer, P.H.; Kutschej, K.; Mitterer, C.; Kathrein, M. A new low friction concept for high temperatures: lubricious oxide formation on sputtered VN coatings. *Tribol. Lett.* **2004**, *17*, 751–756. [CrossRef]
34. Erdemir, A. A crystal-chemical approach to lubrication by solid oxides. *Tribol. Lett.* **2000**, *8*, 97–102. [CrossRef]

Article

Tribological Properties of New Cu-Al/MoS_2 Solid Lubricant Coatings Using Magnetron Sputter Deposition

Ming Cao [1], Lan Zhao [2], Libin Wu [3] and Wenquan Wang [1,*]

1 College of Materials Science and Engineering, Jilin University, No. 5988 Renmin Street, Changchun 130025, China; caoming16@mails.jlu.edu.cn
2 Advanced Materials Research and Development Center, Zhejiang Industry and Trade Vocational College, No. 717 Fudong Road, Wenzhou 325003, China; cm0624@mail.zjitc.net
3 College of Chemistry and Materials Engineering, Wenzhou University, No. 276 Xueyuan Middle Road, Wenzhou 325035, China; 16451282236@wzu.edu.cn
* Correspondence: wwq@jlu.edu.cn; Tel.: +86-136-0440-4468; +86-431-85-094-375

Received: 10 February 2018; Accepted: 26 March 2018; Published: 6 April 2018

Abstract: The increasing demands of environmental protection have led to solid lubricant coatings becoming more and more important. A new type of MoS_2-based coating co-doped with Cu and Al prepared by magnetron sputtering, including Cu/MoS_2 and Cu-Al/MoS_2 coatings, for lubrication applications is reported. To this end, the coatings were annealed in an argon atmosphere furnace. The microstructure and the tribological properties of the coatings prior to and following annealing were analyzed using scanning electron microscopy, energy dispersive spectrometry, X-ray diffractometry (XRD) and with a multi-functional tester for material surface properties. The results demonstrated that the friction coefficient of the Cu/MoS_2 coating was able to reach as low as 0.07, due to the synergistic lubrication effect of the soft metal Cu with MoS_2. However, the wear resistance of the coating was not satisfied. Although the lowest friction coefficient of the Cu-Al/MoS_2 coatings was 0.083, the wear resistance was enhanced, which was attributed to the improved the toughness of the coatings due to the introduction of aluminum. The XRD results revealed that the γ_2-Cu_9Al_4 phase was formed in the specimen of Cu-Al/MoS_2 coatings. The comprehensive performance of the Cu-Al/MoS_2 coatings after annealing was improved in comparison to substrate heating, since the heat-treatment was beneficial for the strengthening of the solid solution of the coatings.

Keywords: Cu/MoS_2 coatings; Cu-Al/MoS_2 coatings; tribological properties; annealing treatment; γ_2-Cu_9Al_4 phase

1. Introduction

The typical materials that can be used as solid lubrication coatings are mainly soft metals (Au, Ag, Cu, Pb), metal compounds (metal oxides, metal halides, metal sulfides, metal selenides), organic materials (Polytetrafluoroetylene, Polyethylene, Polyamide) and et al [1–4]. Among all the metal sulfide systems (FeS, WS_2, MoS_2, ZnS), MoS_2 is widely used, due to its excellent lubricating properties. The excellent lubrication performance of MoS_2 is due to the hexagonal crystal structure of the corresponding layer. MoS_2 exhibits strong chemical bonds within the layers (Mo-S bonds) and weak Van der Waals bonds between individual MoS_2 layers, which cause it to have an anisotropic characteristic. Furthermore, the anisotropy of its crystal structure can improve its load capacity, because the edge planes of MoS_2 have high reactivity and hardness [5]. Conversely, as is commonly known, the usage environment has a significant effect on the performance of MoS_2-based lubricating coatings; for example, the service temperature [6], the humidity [7] and the contact load [4] will all affect the service life of MoS_2-based lubricant coatings. To improve the comprehensive performance of MoS_2-based coatings, a large amount of research has already been conducted. The Physical Vapor Deposition (PVD) method is one of the most efficient methods for preparing MoS_2-based lubricant coatings. Experimental results reveal that metal doping presents a simple and economical method of improving the comprehensive properties of MoS_2-based coatings. It is generally believed that doped soft metals, such as Au [8–10] and Pb [11,12], can reduce the friction coefficients of coatings due to the synergistic lubrication effects, while such coatings have the disadvantage of poor wear resistance. In contrast, the co-doping of metals such as Ti [13,14], Ni [15], Cr [16,17], Al [10,18], Zr [19], Nb [20] and Mo [21] would be appropriate for reducing the friction coefficient and improving the mechanical properties of the coatings. Complementary multi-metal properties can be achieved by using magnetron co-sputtering, which makes up for the deficiency of single-metal. Therefore, there is great potential value in research into the coatings of binary metal co-doped MoS_2. Until recently, few studies had been carried out on the preparation of MoS_2 composite coatings through binary metal co-doping. ILIE [21] studied the MoS_2/(Ti, Mo) composite coatings through Ti-Mo co-doping, and found that the apparent scratches occurred during wear testing, and the overall performance was worse when compared to MoS_2/Ti, indicating that the addition of Mo did not improve the friction properties of the coatings. Ren and co-workers [22] fabricated MoS_2-based coatings by means of the co-sputtering of Pb-Ti, one soft and one hard material, and determined that the surface roughness of the coatings decreased as the Pb content increased. Additionally, the microstructure of cross-section turned from porous columnar into the dense amorphous. The hardness and elastic modulus of the coatings decreased as the Pb content increased, and the tribological properties were improved. Adversely, Pb is an element that causes environmental problems, while the processing of Ag and Au is not easy to control for a high sputtering yield. Since copper and copper alloys are known to be good lubricant additives [23,24], Cu was selected to be the soft metal incorporating into MoS_2, with a view to the practical application of the lubricant coatings. In terms of the strengthening and potential oxidation resistance at higher ambient temperature of the solid solutions, Al was selected as a binary metal for doping into MoS_2. For the purposes of thermal stress reduction and the improvement of the surface uniformity of the coatings, subsequent annealing of the coatings was performed.

Firstly, a series of Cu-MoS_2 coatings was prepared by co-sputtering under various power values of the Cu target. Through the optimization of the process parameters, the Cu-MoS_2 coatings with superior tribological properties were obtained. Subsequently, the co-sputtered Cu-Al/MoS_2 coatings were prepared to further optimize the MoS_2-based coating process. Finally, the prepared composite coatings were annealed under argon atmosphere to obtain superior tribological properties. Through this process, Cu-Al/MoS_2 composite coatings with a new composition system were discovered. The optimal process parameters of the composite coatings were obtained based on the analysis of the relationship between the microstructure and the tribological properties.

2. Materials and Methods

2.1. Preparation of Coatings

The composite lubrication coatings were deposited by a DEER DE500 magnetron sputtering system. A Φ50 mm target of Al (99.999%) and a Φ50 mm target of Cu (99.995%) with direct current (DC) sputtering power, as well as a Φ50 mm target of MoS_2 (99.99%) with radio frequency (RF) power had been used. The coatings were deposited on polished SUS 201 steels and single-crystal silicon wafers synchronously. The test steels of 20 × 20 × 1 mm^3 in dimensions were designed for friction tests. In addition, the steel had an uncoated average roughness (R_a) below 24 nm (mirror finish) and a hardness of 30 HRC. The silicon wafers were used for structural analysis. The substrates were initially cleaned ultrasonically in acetone and an alcohol bath for 20 min, consequently rinsed in deionization water, as well as dried under nitrogen and placed in the vacuum chamber. Prior to coating deposition, an aluminum intermediate layer of approximately 35 nm in thickness was deposited on the substrate.

Then deposition of the main bulk of the MoS_2-based coatings was performed for 3 h. The base pressure in the sputter chamber was about 5×10^{-4} Pa and the working pressure was 1 Pa. Argon was utilized as the working gas for sputtering. The target-substrate distance was 16 cm and the substrate rotation speed was 5 rpm. As presented in Table 1, a series of MoS_2-based coatings was prepared by adjusting the power values of the Cu and Al target, while the power values of the MoS_2 target was retained at 100 W. To improve the bonding force between the substrate and the coating, various substrate temperatures were also utilized in preparing the coatings.

Table 1. Parameters for fabrication of MoS_2-based composite coatings.

Sample No	Target Power/W			Substrate Heat Temp./°C
	MoS_2	Cu	Al	
Cu5	100	5	0	Unheated
Cu10	100	10	0	Unheated
Cu20	100	20	0	Unheated
Cu40	100	40	0	Unheated
Cu50	100	50	0	Unheated
Cu60	100	60	0	Unheated
Cu10(100 °C)	100	10	0	100
Cu10(200 °C)	100	10	0	200
Cu5Al5	100	5	5	Unheated
Cu5Al10	100	5	10	Unheated
Cu10Al5	100	10	5	Unheated
Cu10Al10	100	10	10	Unheated
Cu5Al5(100 °C)	100	5	5	100
Cu5Al5(200 °C)	100	5	5	200

2.2. Annealing of Coatings

It is generally recognized that post-annealing release of residual stress of the coatings [25], without significant phase and lattice constant changes in the as-deposited coatings, even subsequently to annealing under vacuum at higher temperature [26]. Adversely, the results in [27] demonstrated that the MoS_2 layers became crystalline following annealing at approximately 350 °C. In order to observe the effect of annealing temperature on the Cu-Al/MoS_2 composite coatings properties, the coatings were annealed at 300 °C, 400 °C and 500 °C under argon atmosphere based on the phase diagram of Cu-Al [28]. The chamber of controlled atmosphere furnace was evacuated down to a residual pressure of 3 Pa and consequently was purged with argon to remove oxygen. These processes were repeated three times. The argon pressure during annealing was 0.06 MPa. The heating rate was 5 °C/min and the holding time was 2 h. Finally, the coatings were cooled down to the room temperature while the furnace was turned off. The microstructures and tribological properties of the coatings along with subsequent annealing at various temperatures were analyzed and compared.

2.3. Structure Characterization

The 3D-reconstruction of the surface roughness of the composite coatings was obtained with a scanning electron microscope (SEM; Phenom XL, Phenom-World B.V., Eindhoven, The Netherlands). Both the microstructure and cross-sectional morphology of the coatings were observed using a field emission scanning electron microscopy (FESEM; Hitachis-4800, Hitachi Co. Ltd., Tokyo, Japan). The elemental compositions of the coatings were characterized with an X-ray energy-dispersive spectrometer (EDS; Horiba EX250, Horiba Co. Ltd., Kyoto, Japan). The phase structures of the coatings were examined by an X-ray diffractometer (XRD; Bruker D8 Advance, Bruker Inc., Karlsruhe, Germany) with Cu Kα radiation. The scanning range was from 10° to 65° and the scanning rate was 5°/min.

2.4. Tribological Experiments

A reciprocating friction and wear tester (MFT-4000, HuaHui Instruments technology Co. Ltd., Lanzhou, China) was utilized in the tribological testing of the coatings with a ball-on-plate configuration. The friction and wear characteristics of the coatings were evaluated at 22 ± 2 °C, while the relative humidity in testing was approximately 60 ± 10%. The tribological testing were carried out with 2 Hz reciprocating frequency, 5 mm stroke length, 10 N normal forces and a GCr15 steel ball (Φ 4 mm; hardness HRC60) as the friction pair. The corresponding theoretical initial Hertz contact pressure was 1.24 GPa. The average friction coefficients were automatically recorded with a computer. Subsequently to tribological testing, the SEM and optical microscope (OM; Zeiss Scope A1, Carl Zeiss, Gottingen, Germany) were utilized to observe the morphology of the wear tracks. The adhesion quality of the MoS_2-based coatings to the steel substrate was obtained by use of the MFT-4000 scratch tester with a 0.2 mm tip radius diamond indenter. Testing parameter was a scratch length of 5 mm at a load rate of 100 N/min. The different depth profiles of the wear tracks were examined with a surface profilometer (JB-4C; Shanghai Taiming Optical Instrument Co. Ltd., Shanghai, China), that was the grinding crack depth along with the grinding crack width was assumed consistent. The wear scar volumes were calculated according to the Archard wear equation [29] by measuring the average cross-sectional area of track profiles and then multiplying for the track length. The result was used to compare wear rates qualitatively for the errors would exist in the measurements of cross-sectional area with the JB-4C type surface profilometer.

3. Results and Discussion

3.1. Coating Composition and Structural Characterization

Figure 1 depicts the surface and cross-sectional micrographs of the composite coatings prepared with various process parameters. Figure 1a presents a pure MoS_2 coating with worm-shaped surface morphology, which was consistent with the results reported in [30]. The cross-section morphology of the MoS_2 coating was lamellar perpendicularly to the surface. Figure 1b–h presents the micrographs of the Cu/MoS_2 coatings prepared using magnetron co-sputtering and the DC power of Cu target ranged from 5 W to 60 W. The surfaces of the Cu40, Cu50 and Cu60 coatings turned into coarse grained shapes, whereas the cross-section displayed irregular hill shapes with an incompact microstructure. The surface morphologies of the Cu5, Cu10 and Cu20 coatings were smooth, mainly displaying low-sized particles of approximately 20–50 nm, whereas the cross-section presented a dense morphology. Adversely, a low amount of grains of approximately 100–300 nm existed on the coating surfaces of the Cu15 and Cu20 coatings. The cross-sectional microstructure of the Cu10 coating was lamellar perpendicularly to the surface, which was more compact compared to the pure MoS_2 coating. The cross-section of the Cu20 coating was densely columnar perpendicularly to the surface. Figure 1i,j depict the surface micrographs of the Cu10 coatings deposited at 100 °C and 200 °C. The surface morphology of the coatings changed from granular to worm-like as the temperature increased from 100 °C to 200 °C, and the cross-sectional structure became looser. The results demonstrated that the appropriate amount of Cu-doping changed the growth mode of MoS_2, which led to the transformation of the coatings from columnar to densely granular. As the DC sputtering power increased, the granular microstructure on the surface grew rapidly and the surface roughness increased. Consequently, the power range of 10 W–20 W was appropriate for preparing a uniform and dense coating. The substrate-heating accelerated the formation of granular microstructure and increased the sputtering rate of Cu under the same sputtering power. The worm-like morphology in the Cu10 (200 °C) coating similarly to the pure MoS_2 coating was observed. Similar results were found in the literature [31]. The results indicated that the increasing of substrate-heating temperature was beneficial to the formation of MoS_2.

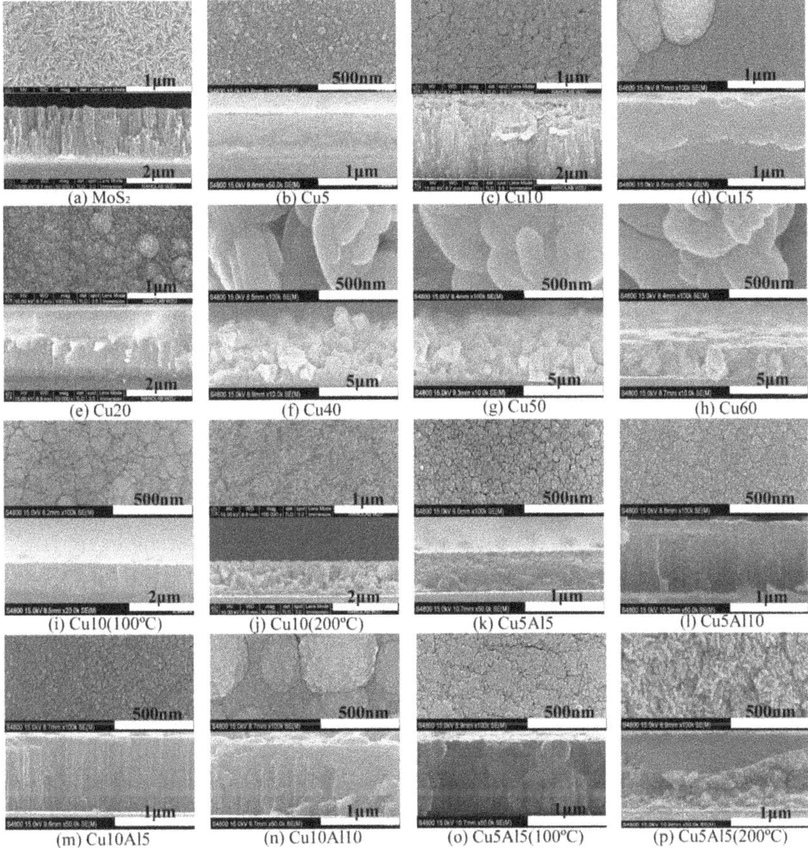

Figure 1. SEM surfaces and cross-sectional morphologies of composite coatings on silicon wafers.

Figure 1k–n presents the surface and cross-sectional micrographs of the coatings, prepared simultaneously using magnetron co-sputtering of the Cu and Al targets. A granular surface morphology with tiny gaps was observed in the Cu5Al5 coating. Figure 1n illustrates certain particles with diameter in the total range of 200–500 nm were generated on the Cu10Al10 coating surface. The cross-sectional morphology of the Cu5Al5 coating was dense granular, whereas the other three coatings were columnar. Figure 1o,p presents the morphologies of the Cu5Al5 composite coatings, as prepared with the substrates heated to 100 °C and 200 °C. Figure 1o presents the granular surface morphology of the Cu5Al5 (100 °C) coating with apparent agglomeration, and the tiny gaps between the agglomerated particles decreased. Moreover, the surface morphology of the Cu5Al5 (200 °C) coating turned into a worm-like shape, whereas the cross-section demonstrated loose microstructure, as presented in Figure 1p, which was consistent with the Cu10 (200 °C) coating. The microstructure analysis of the bimetallic co-sputtered coatings revealed that Al-doping promoted the dense growth of the coatings. Also, to avoid the high-sized grains on the coating surface, the sputtering power of Al and Cu targets should also be reduced properly.

Figure 2 shows the morphologies of the MoS_2-based coatings prior to and following annealing at 300 °C, 400 °C and 500 °C under argon atmosphere. Figure 2a presents the surface microstructure of the MoS_2 coatings prior to and following annealing. There was no apparent change occurred in the surface morphology subsequently to various annealing temperatures. Figure 2b presents the tendency of single grains to agglomerate on the Cu10 coating, whereas the gaps among agglomerations reduced gradually as the temperature increased to 500 °C. Figure 2c presents that the large grain on the Cu20 coating surface disappeared following annealing at 300 °C. The gaps among the surface grains were gradually reduced and the grain size increased following annealing at 400 °C. The grain size of the coating surface increased significantly after annealing at 500 °C. Figure 2d–f presents the surface morphology of the Cu5Al5 coatings prior to and following annealing at different substrate temperatures. The grain-size and compactness of the Cu5Al5 coating surface increased gradually as the annealing temperature increased. The Cu5Al5 (100 °C) and Cu5Al5 (200 °C) coating surfaces also exhibited good uniformity and compactness as the annealing temperature increased. The surface morphology of the Cu5Al5 (200 °C) was similar to the Cu10 (200 °C) coating, which indicated that similar effects in the Cu-Al/MoS_2 coatings might exist while the substrate was heated.

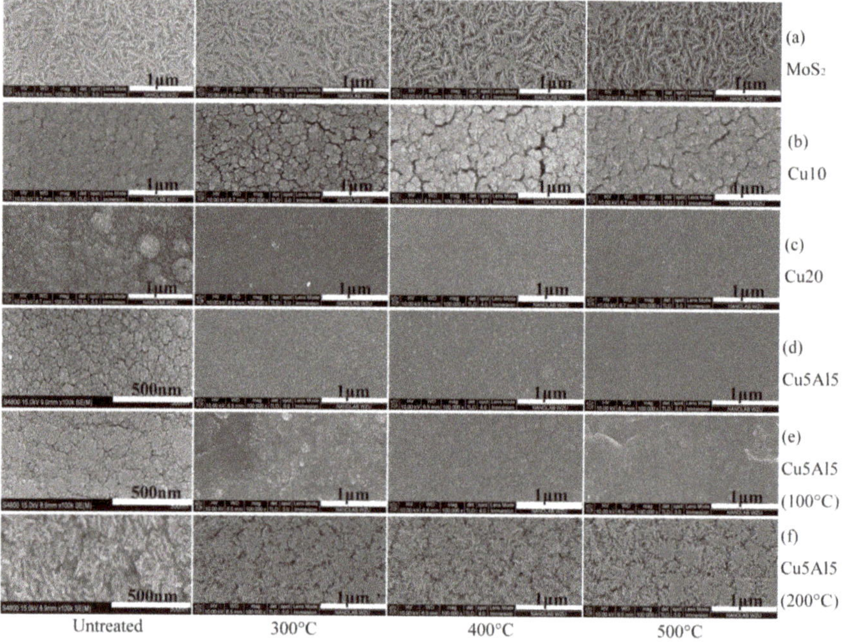

Figure 2. SEM morphologies of composite coatings at various heat treatment temperatures.

Table 2 presents the atomic percentages of the elements in the Cu/MoS_2 coatings. The typical EDS spectrum of Cu10 and Cu20 were shown in Figure 3a,b. The percentage of Cu atoms in the coating increased gradually as the sputtering power increased. The content of Cu in the coating increased from 9.57 at% to 25.13 at%, which indicated the higher sputtering efficiency of Cu within the power range of 10 W–20 W. The ratios of N_S/N_{Mo} in the Cu5–Cu50 coatings were lower than the pure MoS_2 coating, which occurred due to the reverse sputtering of S during the deposition. The reverse sputtering would decrease the deposition rate, which was good for the MoS_2 formation [32]. The value of N_S/N_{Mo} in the Cu5 coating was below 1, whereas fewer S atoms were not conducive to the formation of the MoS_2 (002) basal plane [22] which affected the tribological properties of the coatings. The substrate-heating was beneficial to the deposition of Cu atoms, thereby further affecting the tribological properties of the coatings.

Table 2. Quantitative results of Cu/MoS_2 coatings elemental analysis, determined through EDS.

Composite Coating	Cu/at.%	N_S/N_{Mo}
MoS_2		1.50
Cu5	7.80	0.78
Cu10	9.57	1.18
Cu20	25.13	1.34
Cu40	40.69	1.42
Cu50	46.21	1.50
Cu60	47.99	1.78
Cu10 (100 °C)	28.55	1.30
Cu10 (200 °C)	35.50	1.76

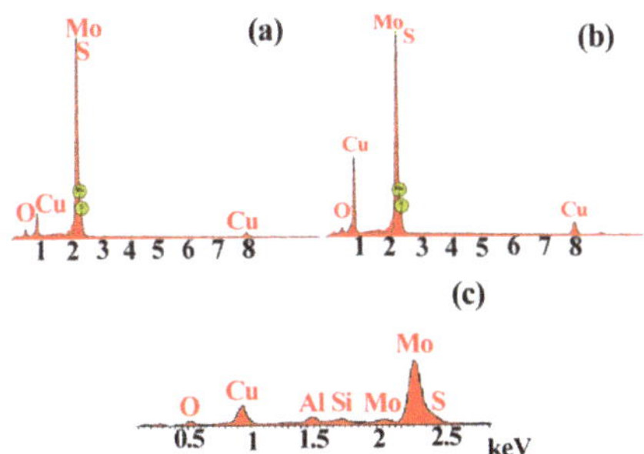

Figure 3. EDS spectrum corresponding to the (**a**) Cu10; (**b**) Cu20; and (**c**) Cu5Al5.

The EDS analysis results of the $Cu\text{-}Al/MoS_2$ coatings are presented in Table 3. The typical EDS spectrum acquired from the area of 10 × 10 μm^2 on the Cu5Al5 coating was shown in Figure 3c. The results indicated that Al-doping promoted the deposition of Cu atoms on the substrate, whereas the substrate heating significantly inhibited the deposition of Cu atoms on the substrate. This might be due to the fact that a competition occurred as a result of partial substitution of Al with Cu. Being the coatings might form a solid solution or an intermediate phase during sputtering and annealing [33]. According to [34], the α-Cu phase would preferentially be produced in the MoS_2-based coatings during the co-sputtering of Cu and Al, followed by a $\gamma_2\text{-}Cu_9Al_4$ intermediate phase with the minimum activation energy. In addition, the N_{Cu}/N_{Al} value of the Cu5Al5 coating shown in Table 3 was the most consistent with the $\gamma_2\text{-}Cu_9Al_4$ intermediate phase. Another consideration is the strength and hardness of the $Cu\text{-}Al/MoS_2$ coatings might increase, whereas the brittleness of the coatings might also be caused by the new phase.

Table 3. Quantitative results of $Cu\text{-}Al/MoS_2$ composite coatings elemental analysis, determined through EDS.

Composite Coating	Cu/at.%	Al/at.%	N_{Cu}/N_{Al}	N_S/N_{Mo}
MoS_2				1.5
Cu5Al5	24.53	10.97	2.24	1.54
Cu5Al10	16.26	8.78	1.85	1.92
Cu10Al5	28.05	4.07	6.88	1.91
Cu10Al10	27.07	5.98	4.52	1.72
Cu5Al5 (100 °C)	10.95	5.63	1.94	1.39
Cu5Al5 (200 °C)	11.42	3.44	3.32	1.58

Figure 4a presents the X-ray diffraction patterns of the Cu/MoS_2 coatings. It demonstrated that the copper phase increased gradually as the Cu target power increased. The crystallization peak of the coatings prepared at low sputtering power of the Cu target was weak, and an amorphous-based structure of the coatings could be deduced, which was consistent with the result found in [22]. Figure 4b presents the XRD patterns of the Cu10 coating subsequently to annealing and the Cu10 coating with the substrate heating. It could be observed that the annealing temperature had an apparent effect on the crystalline of the coatings. When the substrate was heated to 200 °C, the crystallization peak between 10° and 15° was enhanced. It meant that substrate heating was in favor of the crystallization of MoS_2, and MoS_2 tent to be formed with the (002) basal plane parallel to the surface. The XRD patterns of the annealed coatings also proved that annealing could be more beneficial to enhance the crystallization of the coatings than substrate heating.

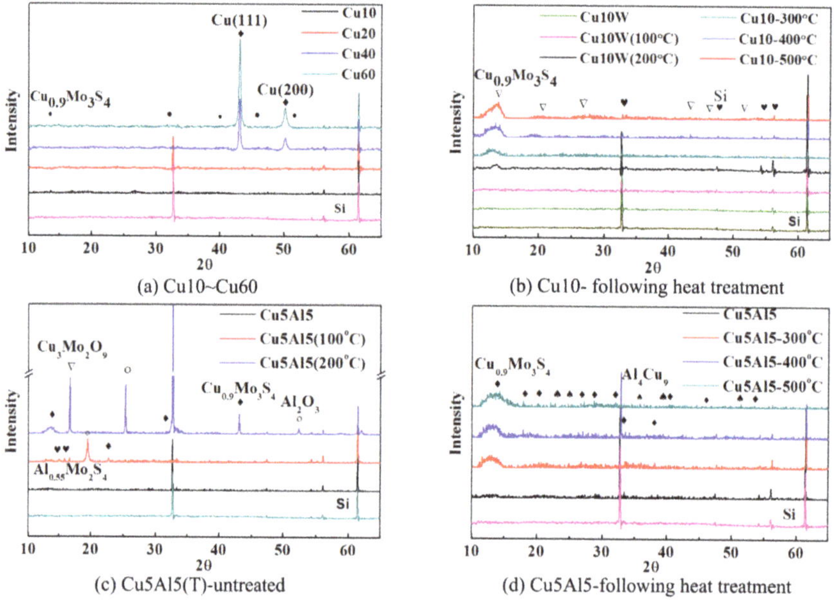

Figure 4. XRD spectrum of Cu/MoS_2 and Cu-Al/MoS_2 composite coatings.

Figure 4c,d presents the XRD patterns of the Cu5Al5 coating with substrate heating and the Cu5Al5 coating following annealing. To a certain extent, the Cu and Al co-doping into MoS_2 inhibited the formation of the MoS_2 crystal structure due to the formation of solid solution or intermediate phase. Simultaneously, the XRD peaks of aluminum phase were not apparent, caused by the Al solid solubility not being exceeded and the overlapping with the MoS2 (002) peak at $2\theta = 13°$ in Figure 4c [18]. In addition, the XRD patterns of the Cu5Al5 (200 °C) coating also revealed the apparent diffraction peaks of $Cu_3MoO_2O_9$ and Al_2O_3. Figure 4d presents the XRD patterns of the Cu5Al5 coating annealed under argon atmosphere, which demonstrated that the γ_2-Cu_9Al_4 phase was formed in the coatings, as reported in [34]. The results indicated that the annealing had beneficial effects on the formation of mesophase in the Cu-Al/MoS_2 coatings.

3.2. Friction and Wear Evaluation

Figure 5 presents the friction coefficient curves of the Cu/MoS_2 coatings. The friction coefficient decreased firstly then increased as the Cu target power increased, as presented in Figure 5a. The friction coefficients of the Cu10 and Cu20 were even lower compared to the pure MoS_2 coating. The average friction coefficient of the Cu10 coating was 0.07, which was the lowest among the Cu/MoS_2 coatings. In Figure 5b, the friction coefficient of the Cu10 (200 °C) was 0.08, which was lower compared to the Cu10 (100 °C), whereas both were higher than the Cu10 coating. The soft metal Cu doping into MoS_2 could reduce the friction coefficient of the MoS_2-based composite coatings. The EDS results demonstrated that the friction coefficient of the Cu10 coating with the N_S/N_{Mo} ratio of 1.18 was the lowest, which was consistent with the results found in [35]. It meant that the Cu10 coating was relatively easy to obtain a MoS_2 (002) basal plane, parallel to the substrate. In addition, the morphology of the Cu10 coating with smooth and dense surface, dense cross-sectional lamellar mentioned above were beneficial to maintain the low friction coefficient of the coatings in the humid environment based on the [36]. In Figure 5c,d, the Cu10 and Cu20 coatings annealed under argon atmosphere displayed an overall friction coefficient increase. When the heat treatment temperature increased to 500 °C, the Cu10 coating had the lowest average friction coefficient (0.088). Moreover, the minimum average friction coefficient of the Cu20 coating was 0.074, when the heat treatment temperature increased to 400 °C. The results indicated that the annealing had more influence on the friction coefficient than substrate heating. The average friction coefficients were reduced as the annealing temperature increased in general.

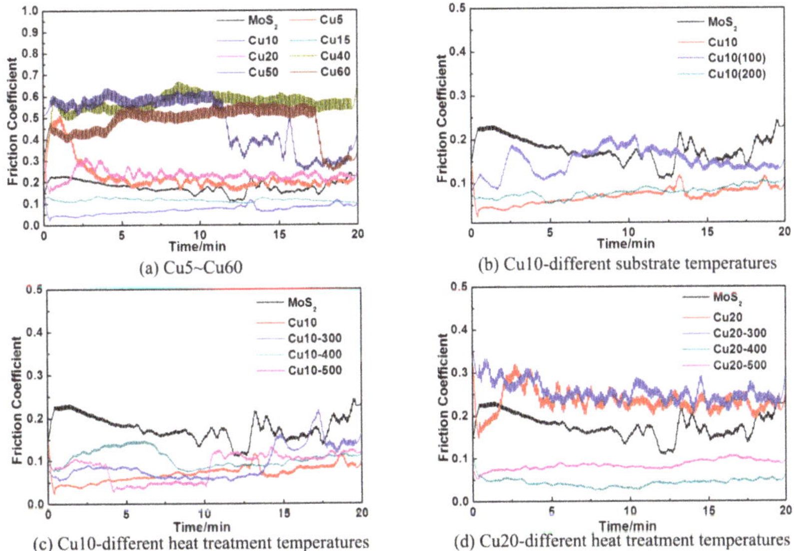

Figure 5. Friction coefficients of Cu/MoS_2 composite coatings at various heat treatment temperatures.

Considering that the friction coefficient of the Cu5Al5 coating (0.083) was the lowest among the Cu-Al/MoS_2 coatings, only the friction coefficient curves of the Cu5Al5 coatings treated at various temperatures were displayed in Figure 6. Figure 6a presents the friction coefficients of the Cu5Al5 coatings at various substrate heat-treatment temperatures and the friction coefficient of Cu5Al5 (100 °C) coating had a maximum value of 0.164. The friction coefficients, as presented in Figure 6b–d, decreased as the heat treatment temperature increased. The friction coefficient of the Cu5Al5 coating annealed at 400 °C was minimum (0.076), where the Cu5Al5 (100 °C) had the lowest friction coefficient (0.068) at the same temperature. The friction coefficient of the Cu5Al5 (200 °C) coating annealed at 500 °C had the lowest value of 0.075. Similar to the Cu/MoS_2 coatings, the annealing was more effective than the substrate heating on the reduction of friction coefficient. Above all, the recommended heat treatment temperature was generally in-between 400 °C and 500 °C.

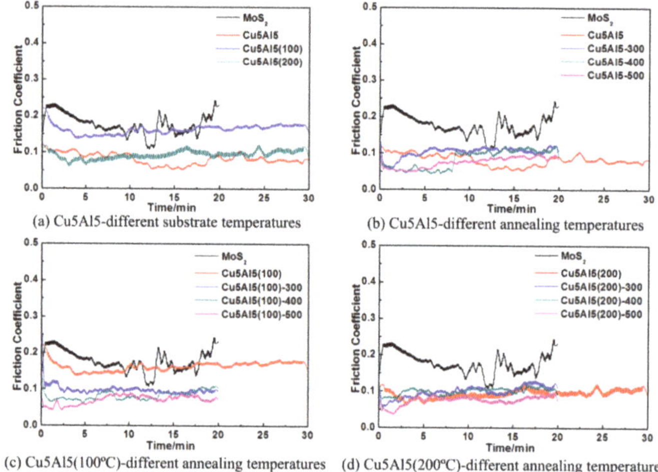

(a) Cu5Al5-different substrate temperatures (b) Cu5Al5-different annealing temperatures (c) Cu5Al5(100°C)-different annealing temperatures (d) Cu5Al5(200°C)-different annealing temperatures

Figure 6. Friction coefficients of the Cu5Al5 composite coatings at various heat treatment temperatures.

Figure 7 presents the typical OM micrographs of the wear tracks o the MoS_2-based coatings. As presented in Figure 7a,e, the wear debris was pushed to the edges of the wear tracks, and there were obvious furrows along with wear tracks on the Cu/MoS_2 coatings. Following annealing, the higher wear resistance of the Cu/MoS_2 coatings would be inferred by the OM micrographs. The results might be caused by a higher density of the coatings according to the SEM morphologies in Figure 2. The wear track of the Cu5Al5 coating, as presented in Figure 7i, was relatively narrower compared to the Cu/MoS_2 coatings, whereas a low amount of wear debris accompanied with shallow furrows existed along the sliding direction. This indicated that the wear degree was significantly lower compared to the Cu/MoS_2 coatings. It also illustrated that the co-doping of Al and Cu could be conducive to the abrasion resistance of the coatings for the formation of the new mesophase. The worn surfaces presented in Figure 7 were covered with grooves parallel to the sliding direction, which were typical features of the abrasive wear.

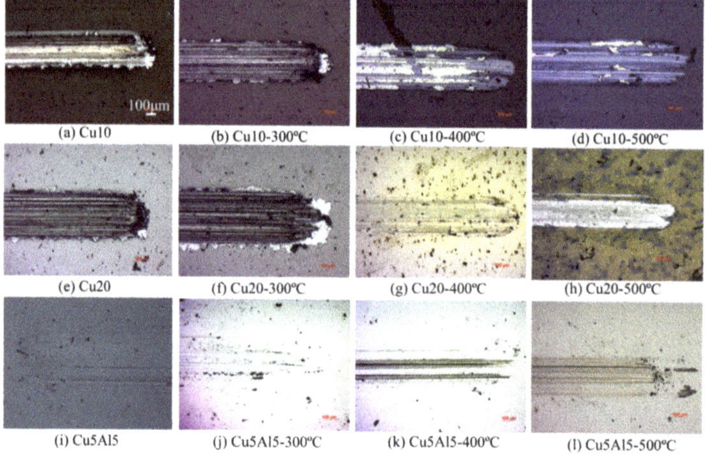

(a) Cu10 (b) Cu10-300°C (c) Cu10-400°C (d) Cu10-500°C (e) Cu20 (f) Cu20-300°C (g) Cu20-400°C (h) Cu20-500°C (i) Cu5Al5 (j) Cu5Al5-300°C (k) Cu5Al5-400°C (l) Cu5Al5-500°C

Figure 7. OM micrographs of MoS_2-based composite coating wear tracks following heat treatment.

3.3. Wear Mechanisms

Scratch testing can be used to evaluate the mechanical failure modes of the coatings. Figure 8 demonstrated the features of gross spallation in the Cu10 coatings case (Figure 8a), brittle tensile cracking followed by gross spallation in the Cu20 coatings case (Figure 8b), conformal cracking followed by gross spallation in the Cu5Al5 (100 °C) coatings case (Figure 8d), conformal cracking followed by bulking spallation in the Cu5Al5 coatings case (Figure 8c) and the Cu5Al5 (200 °C) coatings case (Figure 8e). The hardness of coatings with the tensile cracking followed by spallation was generally considered higher than the coating with conformal cracking followed by spallation and buckling failure, which meant that the Cu20 coatings had a higher hardness. Bulking spallation in the Cu5Al5 and Cu5Al5 (200 °C) coatings meant that the MoS_2 co-doping with Cu and Al improved the toughness of the coatings. The typical wear mechanism of these coatings was the adhesive wear from the results presented in Figure 8.

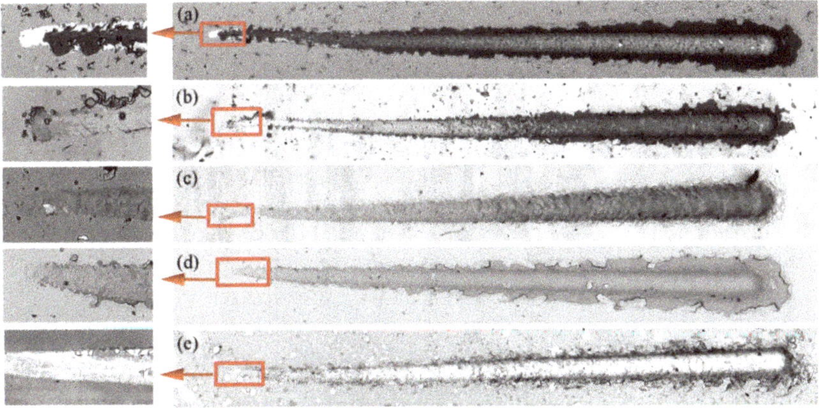

Figure 8. Scratches of (**a**) Cu10; (**b**) Cu20; (**c**) Cu5Al5; (**d**) Cu5Al5 (100 °C); (**e**) Cu5Al5 (200 °C) coatings annealed at 500 °C.

The SEM morphologies and 3D reconstructions of the worn surfaces are presented in Figure 9. Adhesive traces were observed on the worn surfaces of the Cu10 and Cu5Al5 (200 °C) coatings (Figure 9a,d). It was clear that these severe damages resulted from the shearing and rupture of the adhesive junctions between two contacting surfaces. The Cu phase on the friction surface of the Cu10 coating might be softened by braking, which could generate adhesion and lead to material transfer (Figure 9a). The SEM micrographs of typical worn surfaces demonstrated negligible wear scar in the Cu20 coating (Figure 9b), which might be benefit for the enhancement of surface hardness as mentioned above.

The deeper grooves caused by the hard asperities in Figure 9c indicated a primary mechanism of abrasive wear. Hence, a significantly low amount of oxide patches might exist in wear traces according to the EDS analysis (Figure 10a). The EDS results of the Cu5Al5 coating in Figure 10b indicated that a small amount of iron filings produced, while the content of oxygen was less than the Cu10 coating (Figure 10a). Therefore, an oxidation-abrasion mechanism could occur during tribo-testing of the coatings. The worn surface of the Cu5Al5 coating was characterized by typical features of adhesive and abrasive wear, considering the results in Figure 8c. The iron filings or oxide patches on the friction surface could lead to severe wear and deep grooves on the surface.

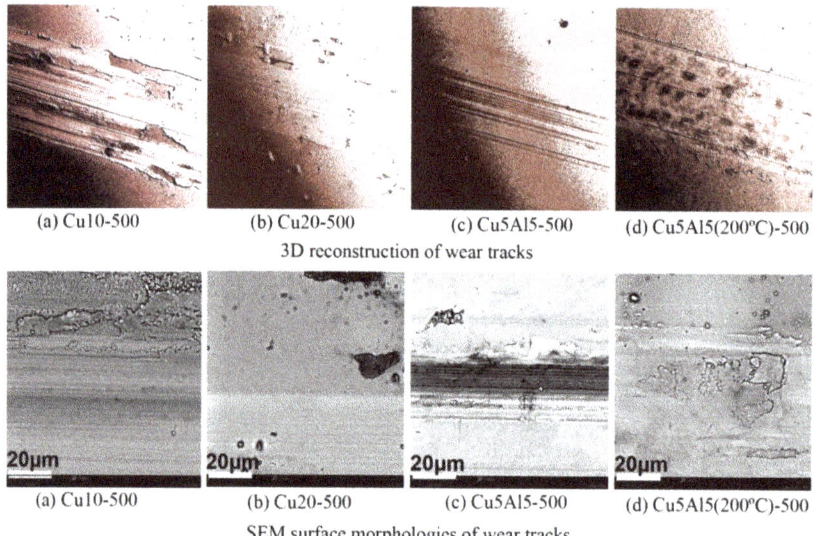

Figure 9. 3D reconstructions and SEM surface morphologies of wear tracks in coatings annealed at 500 °C.

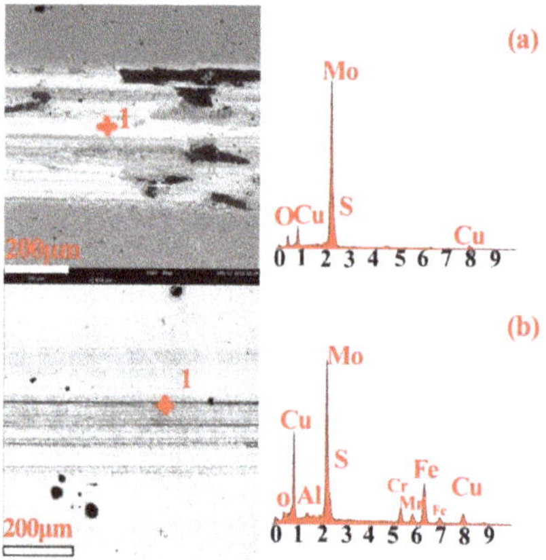

Figure 10. EDS analysis of worn surfaces. (**a**) Cu10; (**b**) Cu5Al5 annealed at 500 °C.

3.4. Oxidation Resistance and Wear Resistance

The oxidation resistance was enhanced according to the EDS testing results on the wear tracks as presented in Figure 11. The Cu5Al5/MoS_2 coating had the best antioxidant properties among all coatings, especially following the annealing at 500 °C. This was because the mesophase promoted the densification of the coatings, and then improved the abrasion resistance and increased the anti-oxidation ability of the coatings during wear tests.

Figure 12 presents a qualitative comparison of the wear rate of MoS_2-based coatings prior to and following annealing. The results showed that both the pure MoS_2 and Cu/MoS_2 coatings had weak wear resistance. By doping soft metal Cu with MoS_2, the friction coefficient of the coatings decreased due to the synergistic lubricating effects. The wear resistance of the Cu/MoS_2 coating was not satisfactory without heat treatment. By analyzing the oxidation potential of Cu, namely the

$\phi^{\theta}(Cu^{2+}/Cu^{+}) = 0.153$ V and $\phi^{\theta}(Cu^{2+}/Cu) = 0.337$ V, the oxygen absorption performance of Cu during wear test was weak. Following annealing, the wear rates of the Cu/MoS_2 coatings began to decrease, especially subsequently to annealing at 500 °C. The SEM micrographs indicated that the dense surface of the coating might be the key factors that affected antioxidant properties.

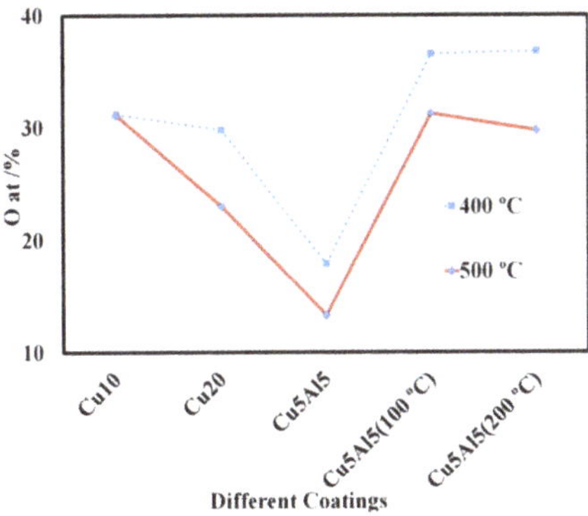

Figure 11. Atomic percentage of oxygen in wear tracks of the coatings annealed at 400 °C and 500 °C.

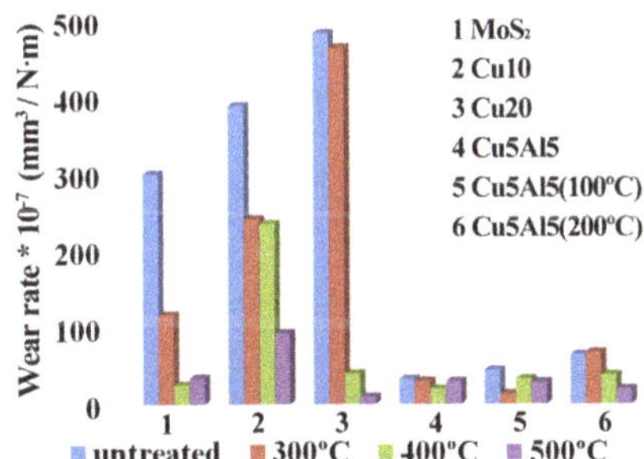

Figure 12. Wear rate of MoS_2-based coatings at various heat treatment temperatures.

The abrasion resistance of the coatings co-doped with Cu and Al was significantly enhanced. On the one hand, the addition of aluminum improved the toughness of the coatings; and on the other hand, the formation of Cu-Al mesophase had contributed to the antioxidant property and moisture resistance of the Cu-Al/MoS_2 coatings. The wear life of the Cu-Al/MoS_2 coatings was thereby improved. Given the oxidation potential of $\phi^{\theta}(Al^{3+}/Al) = -1.63$ V, therefore, Al in the coatings oxidize to Al_2O_3 was easier than Cu. The protective Al_2O_3 layer would help to inhibit the oxygen further infiltrate into the coating matrix. Following annealing at 400 °C, the wear rate of the Cu5Al5 coating decreased to be the lowest. The loose worm-like surface morphology of the Cu5Al5 (200 °C) coating led to an increasing wear rate, but subsequently to an annealing at 500 °C, the wear rate decreased.

4. Conclusions

- The Cu/MoS_2 composite coatings with various Cu contents were prepared by magnetron sputtering and the tribological properties were tested at room temperature. The Cu10 coating had the lowest friction coefficient among all Cu/MoS_2 coatings, whereas the friction coefficient of the Cu5Al5 coating was the lowest in all Cu-Al/MoS_2 coatings. The average friction coefficients of the Cu10 and the Cu5Al5 coatings were both lower than the pure MoS_2 coating. The abrasion resistance of the coatings co-doped with Cu and Al was significantly enhanced through wear rate qualitative comparison. It was very probable that the γ_2-Cu_9Al_4 phase in the co-sputtered Cu-Al/MoS_2 coatings improved the wear resistance.

- Through comparison and analysis of the microstructures observed by SEM, the Cu5Al5 coating with enhanced tribological properties had a smooth surface and a dense cross-sectional non-columnar microstructure. The co-sputtering deposition with Cu and Al led to a smooth surface on the coating. This would lead to the flaking avoidance of the coatings in the friction experiments as well. Following annealing, the grain growth further promoted the compactness of the Cu5Al5 coatings and the wear resistance improvement all at the same time. That was both the low friction coefficient and low wear rate were maintained under the test conditions.
- The surface morphologies of the Cu/MoS_2 and Cu-Al/MoS_2 coatings turned into a worm-like shape, and the cross-section morphology exhibited a loose microstructure. While the substrate was heated to 200 °C, the friction coefficient and wear rate of the coatings increased. Following annealing at 300 °C, 400 °C and 500 °C, the friction coefficient decreased slightly. It meant that the annealing was more effective than the substrate heating on the friction coefficient reduction. Considering the wear rate, the appropriate range of heat treatment temperature was 400 °C–500 °C.
- The scratch test demonstrated the failure mode of conformal cracking followed by bulking spallation in the Cu-Al/MoS_2 coatings, being different from the gross spallation of the Cu10 coating and tensile cracking followed by gross spallation in the Cu20 coating, which had a higher abrasion resistance for the probable cause that the addition of aluminum improved the toughness of the coatings. The worn surface of the Cu5Al5 coating was characterized by typical features of adhesive and abrasive wear.

Acknowledgments: The authors gratefully acknowledge the financial research support of the Project of New Century 551 Talent Nurturing in Wenzhou (2016 No.30). The authors would also like to thank the Zhejiang Provincial Key Laboratory for Cutting Tools for offering access to their instruments and expertise, the Zhejiang Provincial Key Laboratory of Carbon Materials for providing the sample preparation equipment along with the SEM and XRD facilities, as well as the Wenzhou City Key Laboratory of Materials Forming and Tooling Technology for offering the MFT-4000 device and surface roughness testing equipment.

Author Contributions: M.C. and W.W. conceived and designed the experiments; M.C., L.Z. and L.W. performed the experiments; M.C. analyzed the data; M.C., L.Z. and W.W. cooperatively wrote this article.

Conflicts of Interest: The authors declare no conflict of interest.

References

1. Wang, H.; Xu, B.; Liu, J. *Micro and Nano Sulfide Solid Lubrication*, 1st ed.; Springer-Verlag: Berlin/Heidelberg, Germany, 2012; pp. 1–59. ISBN 978-3-642-23101-8.
2. *Materials and Processes for Surface and Interface Engineering*; NATO ASI Series E, 290; Pauleau, Y., Ed.; Kluwer Academic Publishers: Dordrecht, The Netherlands, 1995; pp. 475–527.
3. Menezes, P.L.; Nosonovsky, M.; Ingole, S.P.; Kailas, S.V.; Lovell, M.R. *Tribology for Scientists and Engineers*; Springer: New York, NY, USA, 2013; ISBN 978-1-4614-1945-7.
4. Scharf, T.W.; Prasad, S.V. Solid lubricants: A review. *J. Mater. Sci.* **2013**, *48*, 511–531. [CrossRef]
5. Spalvins, T. Lubrication with sputtered MoS_2 films: Principles, operation, and limitations. *J. Mater. Eng. Perform.* **1992**, *1*, 347–352. [CrossRef]
6. Yang, J.F.; Jiang, Y.; Hardell, J.; Prakash, B.; Fang, Q.F. Influence of service temperature on tribological characteristics of self-lubricant coatings: A review. *Front. Mater. Sci.* **2013**, *7*, 28–39. [CrossRef]
7. Zhao, X.; Zhang, G.; Wang, L.; Xue, Q. The tribological mechanism of MoS_2 film under different humidity. *Tribol. Lett.* **2017**, *65*, 64–71. [CrossRef]
8. Spalvins, T. Frictional and morphological properties of Au–MoS_2 films sputtered from a compact target. *Thin Solid Films* **1984**, *118*, 375–384. [CrossRef]
9. Goeke, R.S.; Kotula, P.G.; Prasad, S.V.; Scharf, T.W. *Synthesis of MoS_2–Au Nanocomposite Films by Sputter Deposition*; Report No. SAND2012-5081; Department of Energy: Washington, DC, USA, 2012.

10. Holbery, J.D.; Pflueger, E.; Savan, A.; Gerbig, Y.; Luo, Q.; Lewis, D.B.; Munzd, W.-D. Alloying MoS_2 with Al and Au: Structure and tribological performance. *Surf. Coat. Technol.* **2003**, *169*, 716–720. [CrossRef]
11. Li, H.; Zhang, G.; Wang, L. Low humidity-sensitivity of MoS_2/Pb nanocomposite coatings. *Wear* **2016**, *350–351*, 1–9. [CrossRef]
12. Wahl, K.J.; Dunn, D.N.; Singer, I.L. Wear behavior of Pb–Mo–S solid lubricating coatings. *Wear* **1999**, *230*, 175–183. [CrossRef]
13. Zhou, H.; Zheng, J.; Wen, Q.P.; Wan, Z.-H.; Sang, R.-P. The effect of Ti content on the structural and mechanical properties of MoS_2–Ti composite coatings deposited by unbalanced magnetron sputtering system. *Phys. Proced.* **2011**, *18*, 234–239.
14. Renevier, N.M.; Oosterling, H.; König, U.; Dautzenberg, H.; Kim, B.J.; Geppert, L.; Koopmans, F.G.M.; Leopold, J. Performance and limitations of MoS_2/Ti composite coated inserts. *Surf. Coat. Technol.* **2003**, *172*, 13–23. [CrossRef]
15. Mikhailov, S.; Savan, A.; Pflüger, E.; Knoblauchc, L.; Hauertc, R.; Simmondsd, M.; Van Swygenhovend, H. Morphology and tribological properties of metal (oxide)–MoS_2 nanostructured multilayer coatings. *Surf. Coat. Technol.* **1998**, *105*, 175–183. [CrossRef]
16. Renevier, N.M.; Fox, V.C.; Teer, D.G.; Hampshire, J. Coating characteristics and tribological properties of sputter-deposited MoS_2/metal composite coatings deposited by closed field unbalanced magnetron sputter ion plating. *Surf. Coat. Technol.* **2000**, *127*, 24–37. [CrossRef]
17. Kao, W.H. Tribological properties and high speed drilling application of MoS_2–Cr coatings. *Wear* **2005**, *258*, 812–825. [CrossRef]
18. Nainaparampil, J.J.; Phani, A.R.; Krzanowski, J.E.; Zabinski, J.S. Pulsed laser-ablated MoS_2–Al films: Friction and wear in humid conditions. *Surf. Coat. Technol.* **2004**, *187*, 326–335. [CrossRef]
19. Song, W.; Deng, J.; Zhang, H.; Yan, P. Study on cutting forces and experiment of MoS_2/Zr-coated cemented carbide tool. *Int. J. Adv. Manuf. Technol.* **2010**, *49*, 903–909.
20. Arslan, E.; Totik, Y.; Bayrak, O.; Efeoglu, I.; Celik, A. High temperature friction and wear behavior of MoS_2/Nb coating in ambient air. *J. Coat. Technol. Res.* **2010**, *7*, 131–137. [CrossRef]
21. Ilie, F.; Tita, C. Tribological properties of solid lubricant nanocomposite coatings obtained by magnetron sputtered of MoS_2/metal (Ti, Mo) nanoparticles. *Proc. Romainan Acad. Ser. A* **2007**, *8*, 207–211.
22. Ren, S.; Li, H.; Cui, M.; Wang, L.; Pu, J. Functional regulation of Pb–Ti/MoS_2 composite coatings for environmentally adaptive solid lubrication. *Appl. Surf. Sci.* **2017**, *401*, 362–372. [CrossRef]
23. Kumar, R.; Sudarshan, T.S. Self-lubricating composites: Graphite-copper. *Mater. Technol.* **1996**, *11*, 191–194. [CrossRef]
24. Chen, S.Y.; Wang, J.; Liu, Y.J.; Liang, J.; Liu, C.S. Synthesis of new Cu-based self-lubricating composites with great mechanical properties. *J. Compos. Mater.* **2011**, *45*, 51–63. [CrossRef]
25. Li, S.P.; Deng, J.X.; Yan, G.Y.; Zhang, K.D.; Zhang, G.D. Microstructure, mechanical properties and tribological performance of TiSiN-WS_2, hard-lubricant coatings. *Appl. Surf. Sci.* **2014**, *309*, 209–217. [CrossRef]
26. Chang, H.W.; Huang, P.K.; Yeh, J.W.; Davison, A.; Tsau, C.H.; Yang, C.C. Influence of substrate bias, deposition temperature and post-deposition annealing on the structure and properties of multi-principal-component (AlCrMoSiTi)N coatings. *Surf. Coat. Technol.* **2008**, *202*, 3360–3366. [CrossRef]
27. Bolster, R.N.; Singe, I.L.; Wegand, J.C.; Fayeulle, S.; Gossett, C.R. Preparation by ion-beam-assisted deposition, analysis and tribological behavior of MoS_2 films. *Surf. Coat. Technol.* **1991**, *46*, 207–216. [CrossRef]
28. Vandenberg, J.M.; Hamm, R.A. An in situ X-ray study of phase formation in Cu–Al thin film couples. *Thin Solid Films* **1982**, *97*, 313–323. [CrossRef]
29. Ramalhoa, A.; Miranda, J.C. The relationship between wear and dissipated energy in sliding systems. *Wear* **2006**, *260*, 361–367. [CrossRef]
30. Buck, V. Morphological properties of sputtered MoS_2 films. *Wear* **1983**, *91*, 281–288. [CrossRef]
31. Lian, Y.S.; Deng, J.X.; Li, S.P. Influence of the deposition temperature on the properties of medium-frequency magnetron sputtering WS_2 soft coated tools. *Adv. Mater. Res.* **2012**, *472–475*, 44–49. [CrossRef]
32. Teresiak, A. Structure investigations of thin MoS_x films. *Mikrochim. Acta* **1997**, *125*, 349–353. [CrossRef]
33. Chen, L.; Paulitsch, J.; Du, Y.; Mayrhofer, P.H. Thermal stability and oxidation resistance of Ti–Al–N coatings. *Surf. Coat. Technol.* **2012**, *206*, 2954–2960. [CrossRef] [PubMed]

34. Wang, W.; Lu, K. Solid state reaction for magnetron sputtering Cu/Al multilayers. *Acta Metall. Sin.* **2003**, *39*, 1–4. (In Chinese)
35. Zhang, X.L.; Hu, N.-S.; He, J.-W. Study on the methods to improve the properties of sputtering deposited MoS_2 films. *J. Mater. Eng.* **1999**, *10*, 44–47. (In Chinese)
36. Fleischauer, P.D. Effects of crystallite orientation on environmental stability and lubrication properties of sputtered MoS_2 thin films. *ASLE Trans.* **1984**, *27*, 82–88. [CrossRef]

Article

Friction and Wear Behavior of an Ag–Mo Co-Implanted GH4169 Alloy via Ion-Beam-Assisted Bombardment

Jiajun Zhu *, Meng Xu, Wulin Yang, Deyi Li, Lingping Zhou and Licai Fu *

College of Materials Science and Engineering, Hunan University, Changsha 410082, China; xumeng@hnu.edu.cn (M.X.); hnuywl@hnu.edu.cn (W.Y.); lideyi@hnu.edu.cn (D.L.); lpzhou@hnu.edu.cn (L.Z.)

* Correspondence: tft@hnu.edu.cn (J.Z.); lfu@hnu.edu.cn (L.F.); Tel./Fax: +86-731-8882-2663 (J.Z. & L.F.)

Academic Editors: Braham Prakash and Jens Hardell

Received: 12 September 2017; Accepted: 2 November 2017; Published: 6 November 2017

Abstract: Ag, Mo, and Ag–Mo were respectively implanted into GH4169 alloy substrates without heating via ion-beam-assisted bombardment technology (IBAB). In addition, the wear performance under low sliding speed and applied load were researched at room temperature (RT). A small amount silver molybdate phase could be detected on the surface of the co-implanted GH4169 alloy bombarded by a high-energy ion beam. The average friction coefficients under the steady wear state had almost no change at all. Compared with the un-implanted GH4169 alloys, the wear rate of the GH4169 alloys with co-implantation of Ag and Mo was reduced by 75%. A large amount of the silver molybdate phase could be generated due to the tribo-reaction on the worn surface during sliding. It benefits the formation of continuous oxide layers as lubrication and protected layers, leading to the change in the predominant wear mechanism from abrasion and adhesion wear to oxidation wear.

Keywords: GH4169 alloy; Ag–Mo co-implantation; ion-beam-assisted bombardment technology (IBAB); wear mechanism

1. Introduction

In order to achieve a low-friction coefficient and wear rate for some harsh-condition applications, numerous works on the lubrication effects about the addition of solid lubricant phases to metal matrices, such as graphite, oxides, soft metals, MoS_2, and fluoride, have been carried out [1–7]. Double metal oxides, including titanates, tungstates, molybdate, and vanadates, are effective high-temperature solid lubricants [8–11]. Among these double metal oxides, more attention has been paid to silver molybdate, which possesses a layer-like structure, because of its excellent lubrication performance, achieving low friction at high temperatures [2,9–17]. For example, Liu et al. [11] investigated the tribological properties of Ni-based self-lubricating composites containing Ag and MoS_2 within a wide temperature range. The evolution of the friction coefficient from room temperature (RT) to 700 °C ranges from 0.18 to 0.77, and the coefficient decreases down to 0.18 at 700 °C as temperature increases. The wear rate from RT to 700 °C is reduced by an order of magnitude, and the worn surface is covered by a continuous and smooth film at 700 °C. As temperature increases to 500 °C, MoS_2 is oxidized and becomes molybdenum oxide. Ag and Mo oxide generates silver molybdate, first proposed by Gulbinski [18], good high-temperature lubrication with a layered structure, which may be considered a mixture of AgO and MoO_3 layers separated by a silver layer. In fact, the lubrication mechanism of silver molybdate can be attributed to weaker Ag–O bridging bonds that are more easily broken to form a rich silver lubrication film when subjected to sliding, which is responsible for the low friction at the sliding interface [9]. Ni-based composites with an addition of lubricious Ag_2MoO_4 at various temperatures have also been explored. Due to high-temperature fabrication processing,

Ag_2MoO_4 decomposes but is reproduced in the rubbing process at high temperatures, leading to an improvement in tribological performance. Thus, at 700 °C, the lowest friction coefficient can be observed (about 0.26). Furthermore, Chen [2] and Aouadi [13] studied adaptive NiCrAlY–Ag–Mo coatings and nanocomposite $Mo_2N/MoS_2/Ag$ coatings, respectively. Friction coefficients for both coatings can be lowered to 0.1 at 600 °C due to the formation of silver molybdate on the worn surface during sliding. Due to the large enthalpy of formation (ΔH_f = +57 kJ/mol), silver molybdate usually forms during sliding at high temperatures and works as an effective high-temperature solid lubricant.

In this study, ion-beam-assisted bombardment (IBAB), an unbalanced process, makes it possible to produce intermetallic compounds or intermediate phases at RT [19–21]. In fact, a silver molybdate phase formed via the co-implanted GH4169 alloy during the preparation process in our experiments, but only a small amount was produced, which may be related to the high surface atomic activity of the Ag and Mo particles on the surface. In addition, a series of tribo-reactions promoted the production of more silver phase on the worn surface during the sliding process. Usually, the powder metallurgy process as a high-temperature fabrication method leads to the decomposition of silver molybdate [11]. The generation of silver molybdate at RT during preparation and sliding has considerable significance, but such generation has scarcely been reported. Moreover, the RT lubrication performance of silver molybdate in the form of a film has not been investigated as much.

This work focuses on the tribological properties of a GH4169 alloy co-implanted with Ag and Mo prepared via IBAB at RT. For comparison, a single Ag and a single Mo are implanted into the same alloy, respectively. A correlation analysis and a discussion of the tribo-chemical reaction on the worn surface and effects on the wear behavior are given. The wear mechanism is further revealed as well.

2. Experimental Details

2.1. Sample Preparation

The chemical compositions of the commercial GH4169 alloy, provided by Beijing Iron and Steel Research Institute, are shown in the following Table 1.

Table 1. The chemical compositions of the GH4169 alloy.

Chemical Element	Weight Percentage (wt %)
Fe	20.2
Cr	16.8
Nb	4.95
Mo	2.81
Ti	0.78
Al	0.46
Si	0.49
Ni	balance

The polished 15 mm × 15 mm × 5 mm GH4169 alloy substrates were implanted with Ag, Mo, and Ag–Mo via IBAB. These preparation experiments were performed on a homemade multi-functional ion beam system (MIS800, Sykeyou Vacuum Technology Institute, Shenyang, China), the structure schematic of which is shown in Figure 1. Two Kaufman ion sources (named sputtering ion sources 1 and 2) were respectively used for sputtering an Ag target (purity > 99.99%) and an Mo target (purity > 99.9%) with a 2.5 keV Ar^+ ion beam. A third Kaufman ion source (named the assisted ion source) was used for bombarding the particles into the alloy surface via an Ar^+ ion beam with an accelerating voltage of about several tens of thousands of volts, which was incident along the normal direction of the base surface. The purity of argon as a working gas was about 99.999%, and the working pressures during deposition with single and double targets was set to 1.5×10^{-2} Pa and 2.3×10^{-2} Pa, respectively. The experimental time was 1 h. The implanted samples were numbered in accordance with the implanted particle, as shown in Table 2. The un-implanted alloy was marked as N.

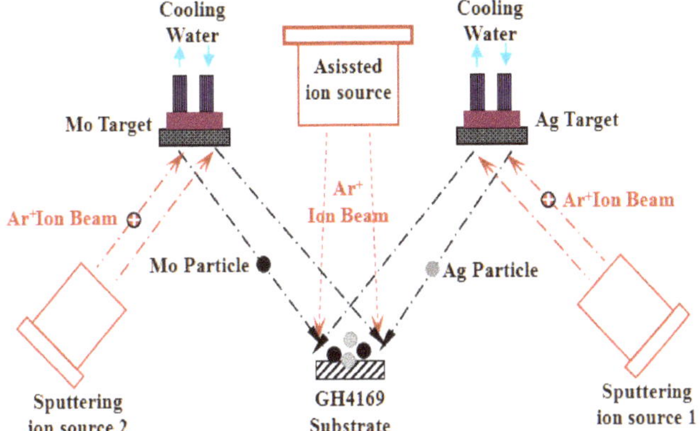

Figure 1. Schematic diagram of the multi-functional ion beam deposition system.

Table 2. The particles implanted into GH4169 alloy substrates and the numbering.

Sample	Particle Implanted into the GH4169 Alloy
A	Silver
M	Molybdenum
AM	Silver and molybdenum

2.2. Friction and Wear Test

The friction coefficients of all samples were measured under a normal load of 2 N in atmospheric conditions at RT with a wear tribometer (WTM-2E, Lanzhou Institute of Chemical Physics, Lanzhou, China) in a ball-on-disk style (a relative humidity of 70% ± 5%). The wear test was conducted in a circular motion 5 mm in diameter under a fixed sliding speed of 47 mm/s. The upper ball was a counterpart Si_3N_4 ball 5 mm in diameter with a hardness of 19 GPa. Hertzian contact pressure was 1.2 GPa, according to our calculations. The single wear test time was 10 min and was repeated five times. The friction coefficient curves were recorded automatically with a computer attached to the tribometer. After the wear test, the Wyko NT9100 surface profiler (Chung Ming Automation Equipment Co., Ltd., Shanghai, China) was used to obtain the profiles of the wear tracks. The wear rate was calculated as the equation $\omega = \Delta V/(F{\cdot}L)$, ω was the wear rate in $mm^3{\cdot}N^{-1}{\cdot}m^{-1}$, ΔV was the wear volume loss in mm^3, L was the total motion distance in m, and F was the load in N.

2.3. Characterization

The structures and the phase compositions of all samples were analyzed with an X-ray diffractometer (XRD, D500, Bruker, Beijing, China) with a 40 kV operating voltage and Cu Kα radiation in the 2θ range of 25°–80°. The chemical states before the friction test were examined with an X-ray photoelectron spectroscope (XPS, PHI-5702, Thermo Fisher Scientific, Waltham, MA, USA). The Raman spectrometer (Labram-010, an excitation wavelength of 632 nm, HORIBA, Beijing, China) further surveyed the phase constitutions of all samples before and after the wear test. The elemental compositions of selected regions and the morphologies of the wear tracks for all samples were analyzed using a scanning electron microscope (SEM, JSM-6700F, JEOL, Beijing, China), equipped with backscattered electrons (BSEs) and an energy dispersive spectrometer (EDS).

3. Result and Discussion

3.1. Composition and Microstructure

The XRD patterns of all samples before the wear test are presented in Figure 2, which includes the un-implanted GH4169 alloy. It was found that the γ matrix phase could be observed in all samples, which mainly came from the GH4169 alloy. In Figure 2ab, the characteristic peaks of Ag (2θ = 37.92°) and Mo (2θ = 40.56°) correspond to FCC Ag (111) and BCC Mo (110), respectively, excluding the GH4169 base. Both Ag (111) and Mo (110) peaks have a smaller angle shift. It may be related to the lattice distortion induced by the bombardment of the high-energy ion beam. Moreover, the XRD results of the co-implanted GH4169 alloy allowed us to identify the Ag and Mo phases in the co-implanted sample (Figure 2d).

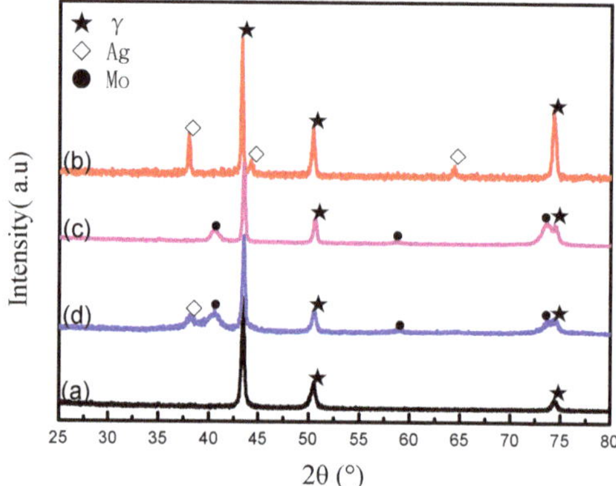

Figure 2. XRD spectra of (**a**) the un-implanted alloy (N); (**b**) silver (A); (**c**) molybdenum (M); and (**d**) silver and molybdenum (AM) before the wear test.

Figure 3a, in which peaks derived from O, Ag, and Mo are all evident, exhibits the XPS full spectrum for the co-implanted GH4169 alloy before the wear test. As can be seen in Figure 3b, the XPS peaks of Ols at 531.8 eV and 530.5 eV yielded two Ag_2MoO_4 Gaussian peaks. The $Ag3d_{5/2}$ and $3d_{3/2}$ peaks deviate from those of the Ag-implanted sample (Figure 3c), indicating that the binding energy of $Ag3d_{5/2}$ and $3d_{3/2}$ slightly shifted to a lower value. Similarly, the $Mo3d_{5/2}$ and $3d_{3/2}$ peaks deviate from those of the Mo-implanted sample (Figure 3d). Based on the Ag3d peak position, the peaks at 368.18 eV and 374.23 eV can be assigned to $Ag3d_{5/2}$ and $3d_{3/2}$ in the Ag_2MoO_4 phase, respectively. In addition, the peaks located at 232.73 eV and 235.93 eV can be assigned to $Mo3d_{5/2}$ and $3d_{3/2}$ in Ag_2MoO_4, respectively. This result is consistent with reports of Suthanthiraraj and Liu [11,22].

Compared with the XPS result above, no obvious characteristic peaks of the Ag_2MoO_4 phase were found in the XRD pattern of AM before sliding. This may be due to the low content and good dispersion of Ag_2MoO_4. To confirm this, the phase compositions of AM were determined via a laser Raman spectrometer (Figure 4). Monoclinic silver molybdate with a broad peak can be detected with Raman spectra [2,11–13]. We inferred that silver molybdate indeed formed during the preparation process. Because of the bombardment of the high-energy ion beam, the co-deposited particles on the base surface have substantially high activity. The chemical reaction between Ag, Mo, and the residual oxygen from the vacuum chamber or the oxygen adsorbed on the base surface might happen under an unbalanced fabrication process at RT. XPS and Raman analysis reveal that silver molybdate can be generated directly via ion-beam-assisted bombardment at RT.

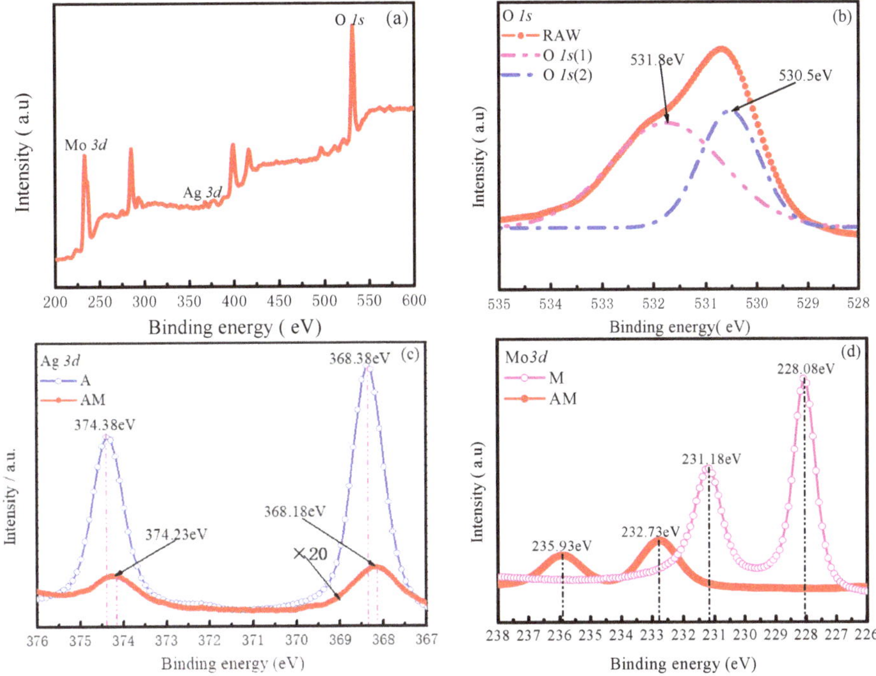

Figure 3. (**a**) XPS full spectra; (**b**) O1s XPS spectra for AM; (**c**) Ag3d XPS spectra of A and AM; and (**d**) Mo3d XPS spectra of M and AM before the wear test.

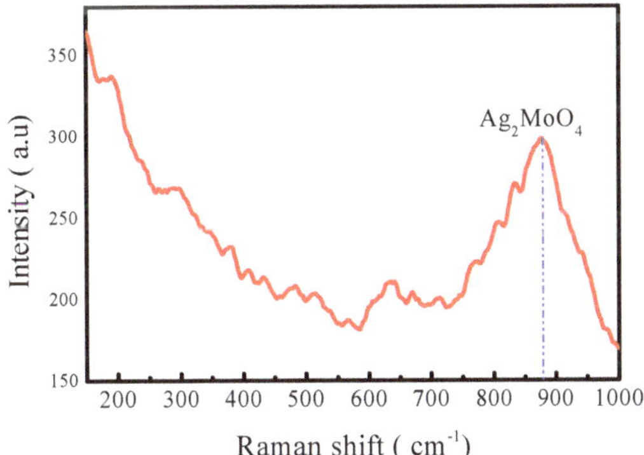

Figure 4. Raman spectra of AM before the wear test.

3.2. Friction and Wear Properties

Figure 5a presents the friction coefficient evolution of all samples with sliding time under a normal load of 2 N. Compared with the bare GH4169 alloy (N), the AM shows a lower friction coefficient at the beginning of the test, which is attributed to the formation of silver molybdate with an atomically layered structure during the wear test. As the test proceeded, the friction coefficient steadily increased to reach the base nature. The average friction coefficient and wear rate under the dry friction and RT conditions are presented in Figure 5b for all samples. Compared with the bare GH4169 alloy, there is no obvious change in the average friction coefficient at the steady wear state. It was speculated that the

Ag-implanted sample has a smaller mean friction coefficient due to the synergistic lubrication effect of Ag and Ag_2O. The wear rate of the co-implanted sample (AM) was the lowest of all samples, as shown in Figure 5b, from 4.88×10^{-4} $mm^3 \cdot N^{-1} \cdot m^{-1}$ to 1.22×10^{-4} $mm^3 \cdot N^{-1} \cdot m^{-1}$ compared with the base, due to the presence of silver molybdate. The wear rate for the Mo-implanted sample is much higher, the highest of all the samples, than that of the base. It is also estimated that MoO_3 formed on the worn surface leads into the worst anti-wear performance for M. Research shows that MoO_3 is brittle and has no lubrication at low temperature [23].

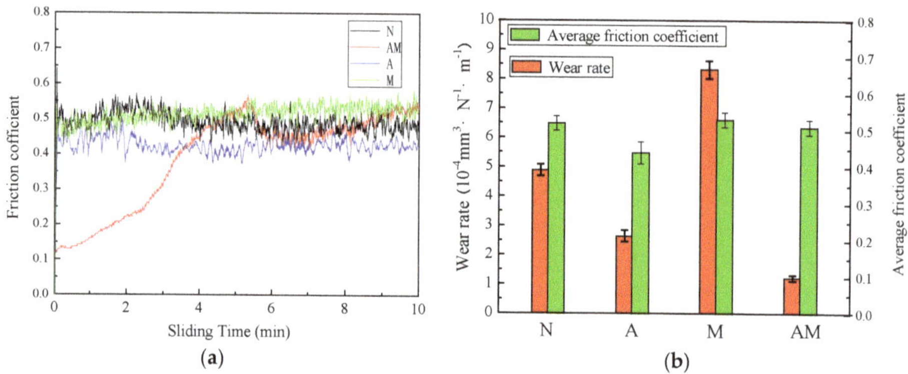

Figure 5. (**a**) The friction coefficient and (**b**) the average friction coefficient and wear rate for all samples tested under a normal load of 2 N and a fixed sliding speed of 47 mm/s.

Figure 6 displays the cross-section profiles and surface 3D topographies of the wear scars for all samples after the 10 min friction test. Among all samples, the sample AM exhibits the smoothest wear tracks with the smallest wear volumes as seen in Figure 6d, indicating the best wear resistance performance. The visible adhesive marks are observed on the selected zone marked with a red box for the worn surface of N, A, and M as seen from Figure 6a–c. The Mo-implanted sample shows the deepest wear tracks (Figure 6c), which is consistent with Figure 5.

The SEM morphologies of the wear tracks after sliding are given for all samples in Figure 7. The selected dark zone (named I) and gray zone (named II) on the wear scars for all samples were analyzed with EDS. The oxygen element was enriched in the dark regions. In Figure 7a, the worn surface is characterized with small bumps and delamination pits, indicating abrasive wear and brittle fracture in the bare GH4169 alloy. The discontinuous dark regions covering the surface in Figure 7b are believed to be the discontinuous oxide layer. It can be inferred that the predominant wear mechanism for A is oxidation and adhesive wear. The Mo-implanted sample, accompanied by a large amount of the wear debris and some abrasive grooves on the worn surface, exhibits the most serious wear condition (Figure 7c). The continuous tribo-layer covered on the worn surface, together with some shallow grooves for the co-implanted sample, can be seen in Figure 7d. As discussed before, the continuous oxide layers (the corresponding continuous dark regions) formed in AM with the smoothest wear scar, which accounts for the lower wear rate. The formation of a tribo-layer, as a protective and lubricating layer [24], can effectively reduce the direct contact of the sample and the Si_3N_4 ball. It has been illustrated that the oxidation wear dominates the wear processing of the co-implanted GH4169 alloy [25].

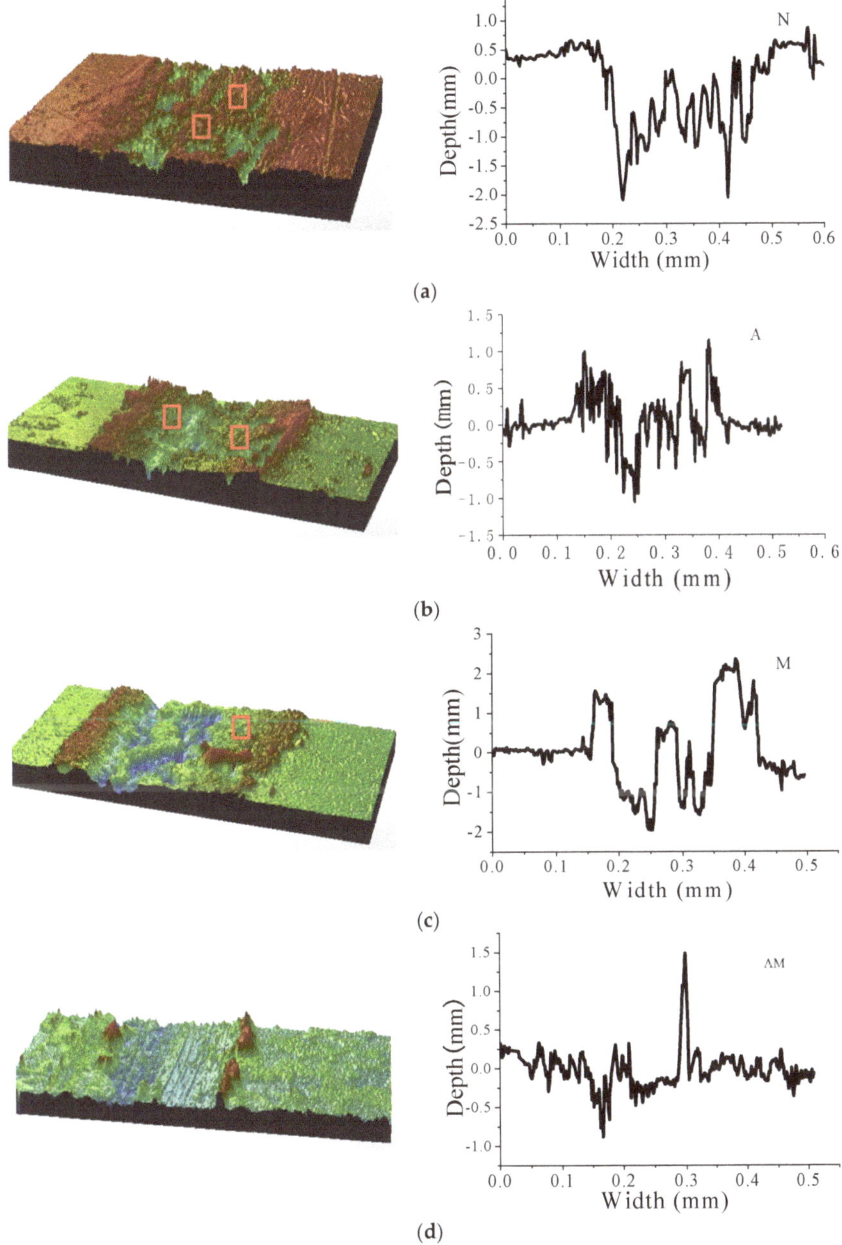

Figure 6. The cross-section profiles, surface 3D topographies of wear tracks for (**a**) N, (**b**) A, (**c**) M, and (**d**) AM after the wear test.

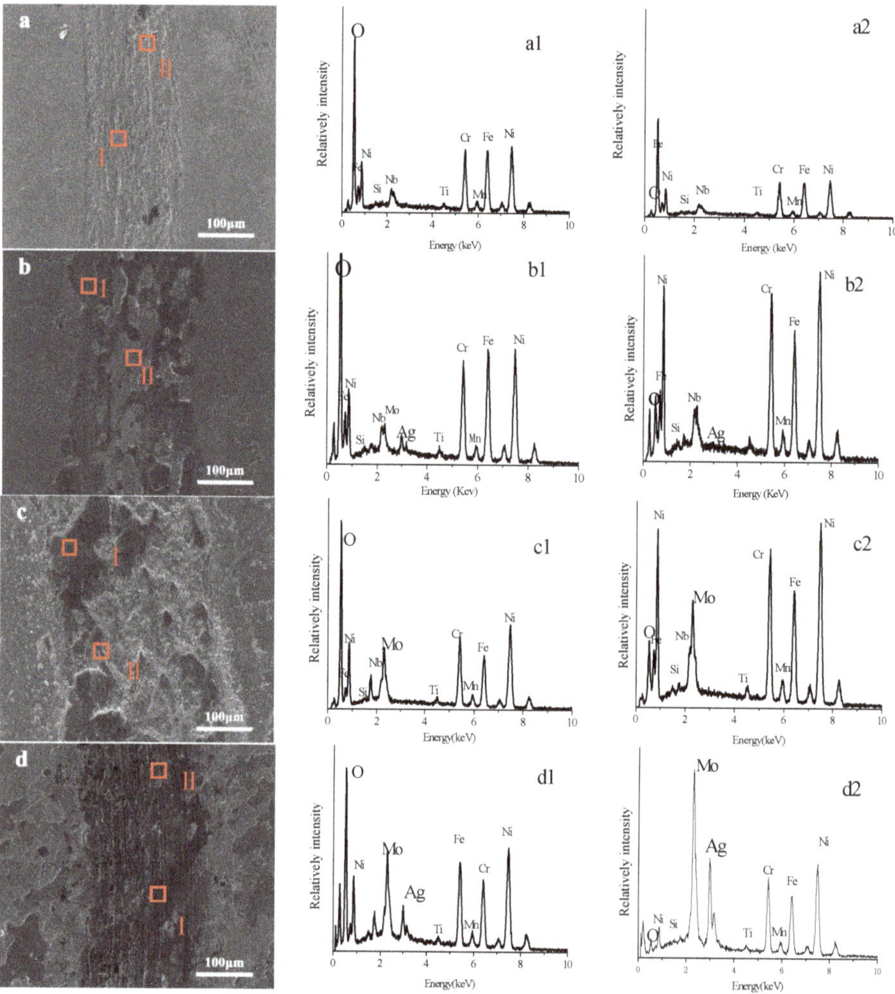

Figure 7. The SEM micrographs of the worn surfaces after the fiction test for (**a**) N, (**b**) A, (**c**) M, and (**d**) AM.

Adhesion wear can be seen in the SEM images in Figure 8 on the worn surface of the counterface ball against the AM. The SEM image of the worn surface of the counterface ball against A presents clear abrasion wear. This is because, as the wear test continued, the contact between the counterface and the sample (A and AM) led to material transfer via adhesive wear. Table 3 shows the chemical compositions of the counterface balls determined via EDS, providing further evidence for oxidation wear and material transfer via adhesion wear.

The tribo-chemical reactions and phase changes on the worn surfaces were further studied with the micro-Raman. In the following Raman spectra (Figure 9a), Fe_2O_3, resulting from the oxidation of Fe and Cr in the matrix, can be detected on the worn surfaces of all samples. The generation of Fe_2O_3 is expected to contribute to the lubricating effect [23,26,27]. Ag_2O and MoO_3 can be found on the worn surfaces, respectively, of A and M. In addition, the Ag_2MoO_4 phase is detected on the worn surface after the sliding process of AM. It has been inferred that silver molybdate can be combined via tribo-reaction between MoO_3 and Ag on high-energy-contacted surfaces during RT wear tests [2,11]. Based on the Raman spectra of the sample surfaces, as shown in Figure 9b, before and after the wear test, the peak intensity of Ag_2MoO_4 increased substantially after the friction test of AM. It can be inferred that Ag_2MoO_4 content rose after the wear test due to the further generation of Ag_2MoO_4 during sliding. That the Ag_2MoO_4 phase can be synthesized during the preparation and rubbing processes is also demonstrated.

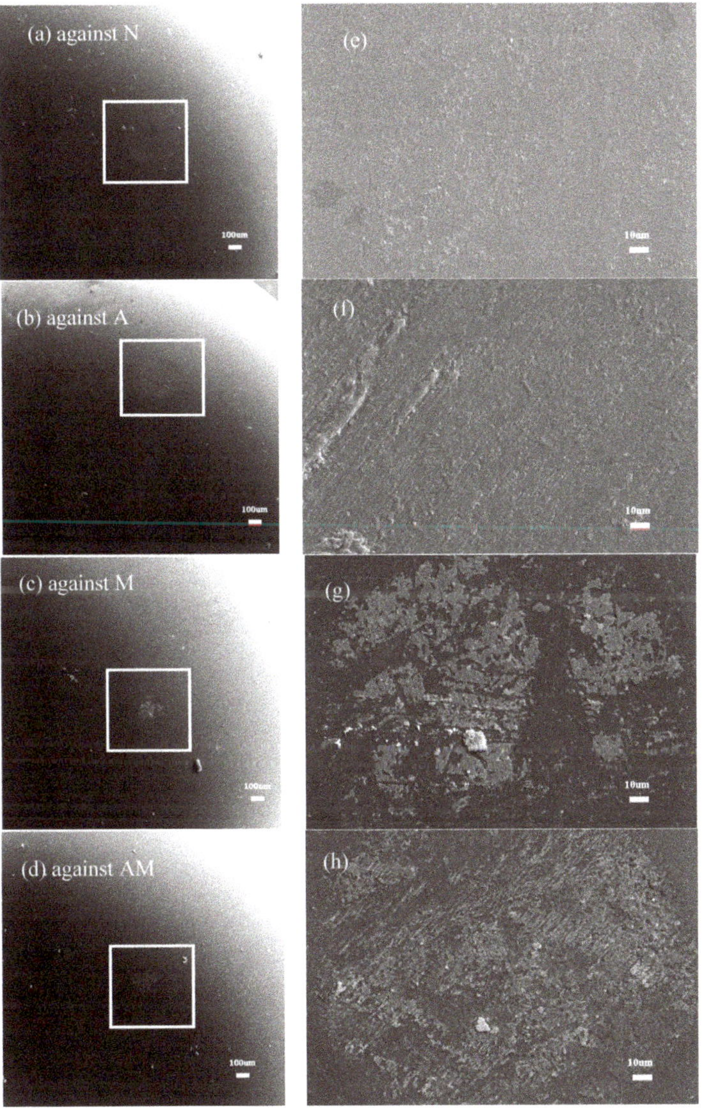

Figure 8. The SEM images of the counterface balls (Si_3N_4 ball) after sliding against (**a**) A, (**b**) B, (**c**) C, and (**d**) D and corresponding higher magnification SEM images against (**e**) A, (**f**) B, (**g**) C, and (**h**) D.

Table 3. The element compositions of the worn surfaces of all counterface balls.

Sample	Element Composition (at.%)							
	Si	N	Ni	Fe	O	Ag	Al	Mo
N	72.96	7.36	0.61	0.17	16.62	0	2.27	0
A	56.54	19.78	0.89	1.27	19.19	0.43	1.91	0
M	74.41	6.21	0.55	0.66	16.29	0	1.87	0
AM	56.14	14.8	0.33	0.55	23.28	0.99	1.98	1.93

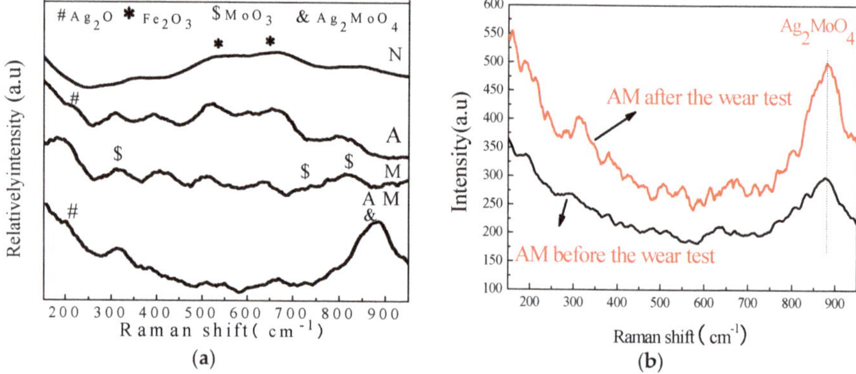

Figure 9. (**a**) Raman spectra of the worn surfaces for all samples after the wear test; and (**b**) Raman spectra before and after the wear test for AM.

As demonstrated above, the unbalanced fabrication method results in the generation of Ag_2MoO_4 during the preparation process at RT. In addition, more silver molybdate formed on the worn surface of AM during the RT friction-wear test through a series of tribo-chemical reactions on the high-energy-contacted surface. The co-implanted sample has the best wear resistance performance, attributed to the formation of Ag_2MoO_4. In the friction test, the wear mechanism was dominant by oxidation wear in AM covered by a continuous oxide layer. A large amount of silver molybdate with a layered structure [11,28], formed on the worn surface during sliding, promotes the formation of a continuous oxide film. Furthermore, the Ag_2MoO_4 phase with weaker Ag–O bridging bonds can easily break to form rich silver lubrication layers at a lower wear rate. It is generally known that silver molybdate is an excellent high-temperature solid lubricant, and little research on the lubrication behavior of Ag_2MoO_4 at RT has been conducted. However, our experimental results prove that silver molybdate can greatly improve the wear resistance performance at RT.

4. Conclusions

By ion-beam-assisted bombardment technology, the GH4169 alloy implanted with Ag, Mo, and Ag–Mo, respectively, were prepared at RT. The microstructure, phase compositions, tribological behavior and mechanism were investigated. The conclusions drawn are as follows:

- Compared with GH4169 alloy substrates, the wear resistance of the GH4169 alloy co-implanted with Ag and Mo was effectively enhanced, which is attributed to the production of Ag_2MoO_4 with a layer-like structure, easily forming a rich silver lubrication layer during sliding.
- The predominant wear mechanism changed from abrasion and adhesion wear to oxidation wear due to the silver lubrication layer and continuous oxide layers found on the worn surface of the co-implanted GH4169 alloy.

- During the preparation process and sliding at RT, the formation of silver molybdate can be achieved through ion-beam-assisted bombardment, an unbalanced preparation method.

Acknowledgments: The authors acknowledge the financial support from the National Natural Science Foundation of China (51401080 and 51675169).

Author Contributions: Jiajun Zhu and Licai Fu conceived and designed the experiments; Lingping Zhou performed the experiments; Jiajun Zhu and Licai Fu analyzed the data; Wulin Yang and Deyi Li contributed reagents/materials/analysis tools; Jiajun Zhu and Meng Xu wrote the paper.

Conflicts of Interest: The authors declare no conflict of interest.

References

1. An, Y.; Chen, J.; Hou, G.; Zhao, X.; Zhou, H.; Chen, J.; Yan, F. Effect of silver content on tribological property and thermal stability of HVOF-sprayed nickel-based solid lubrication coating. *Tribol. Lett.* **2015**, *58*, 34. [CrossRef]
2. Chen, J.; An, Y.; Yang, J.; Zhao, X.; Zhou, H.; Chen, J.; Yan, F. Tribological properties of adaptive NiCrAlY–Ag–Mo coatings prepared by atmospheric plasma spraying. *Surf. Coat. Technol.* **2013**, *235*, 521–528. [CrossRef]
3. Yin, H.; Xu, Y.; Li, X.; Chang, W.; Zhou, Y. Design of friction and wear resistant titanium- and cobalt-modified nickel-base repair alloys by spray forming. *Mater. Des.* **2017**, *116*, 403–410. [CrossRef]
4. Scharf, T.W.; Prasad, S.V. Solid lubricants: A review. *J. Mater. Sci.* **2013**, *48*, 511–531. [CrossRef]
5. Cheng, J.; Zhen, J.; Zhu, S.; Yang, J.; Ma, J.; Li, W.; Liu, W. Friction and wear behavior of Ni-based solid-lubricating composites at high temperature in a vacuum environment. *Mater. Des.* **2017**, *122*, 405–413. [CrossRef]
6. Zhu, S.; Li, F.; Ma, J.; Cheng, J.; Yin, B.; Yang, J. Tribological properties of Ni_3Al matrix composites with addition of silver and barium salt. *Tribol. Int.* **2015**, *84*, 118–123. [CrossRef]
7. Zhu, S.; Bi, Q.; Yang, J.; Liu, W. Effect of fluoride content on friction and wear performance of Ni_3Al matrix high-temperature self-lubricating composites. *Tribol. Lett.* **2011**, *43*, 341–349. [CrossRef]
8. Stone, D.S.; Harbin, S.; Mohseni, H.; Mogonye, J.E.; Scharf, T.W.; Muratore, C. Lubricious silver tantalate films for extreme temperature applications. *Surf. Coat. Technol.* **2013**, *217*, 140–146. [CrossRef]
9. Gao, H.Y.; Otero-De-La-Roza, A.; Gu, J.J.; Stone, D.A.; Aouadi, S.M. (Ag,Cu)–Ta–O ternaries as high-temperature solid-lubricant coatings. *ACS Appl. Mater. Interfaces* **2015**, *7*, 15422–15429. [CrossRef] [PubMed]
10. Liu, E.-Y.; Wang, W.-Z.; Gao, Y.-M.; Jia, J.-H. Tribological properties of adaptive Ni-based composites with addition of lubricious Ag_2MoO_4 at elevated temperatures. *Tribol. Lett.* **2012**, *47*, 21–30. [CrossRef]
11. Liu, E.-Y.; Wang, W.-Z.; Gao, Y.-M.; Jia, J.-H. Tribological properties of Ni-based self-lubricating composites with addition of silver and molybdenum disulfide. *Tribol. Int.* **2013**, *57*, 235–241. [CrossRef]
12. Bondarev, A.V.; Kiryukhantsev-Korneev, P.V.; Sidorenko, D.A.; Shtansky, D.V. A new insight into hard low friction MoCN–Ag coatings intended for applications in wide temperature range. *Mater. Des.* **2016**, *93*, 63–72. [CrossRef]
13. Aouadi, S.M.; Paudel, Y.; Simonson, W.J.; Ge, Q.; Kohli, P.; Muratore, C. Tribological investigation of adaptive $Mo_2N/MoS_2/Ag$ coatings with high sulfur content. *Surf. Coat. Technol.* **2009**, *203*, 1304–1309. [CrossRef]
14. Chen, F.; Feng, Y.; Shao, H.; Zhang, X.; Chen, J.; Chen, N. Friction and Wear Behaviors of $Ag/MoS_2/G$ Composite in Different Atmospheres and at Different Temperatures. *Tribol. Lett.* **2012**, *47*, 139–148. [CrossRef]
15. Gulbiński, W.; Suszko, T. Thin films of MoO_3–Ag_2O binary oxides—The high temperature lubricants. *Wear* **2006**, *261*, 867–873. [CrossRef]
16. Jie, C.; Zhao, X.; Zhou, H. HVOF-sprayed adaptive low friction NiMoAl–Ag coating for tribological application from 20 to 800 °C. *Tribol. Lett.* **2014**, *56*, 55–66.
17. Li, B.; Gao, Y.; Han, M.; Guo, H.; Wang, W.; Jia, J. Microstructure and tribological properties of NiCrAlY-Mo-Ag composite by vacuum hot-press sintering. *Vacuum* **2017**, *143*, 1–6. [CrossRef]
18. Gulbinski, W.; Suszko, T.; Sienicki, W.; Warcholinski, B. Tribological properties of silver- and copper-doped transition metal oxide coatings. *Wear* **2003**, *254*, 129–135. [CrossRef]

19. Figueroa, R.; Abreu, C.M.; Cristóbal, M.J.; Pena, A. Effect of nitrogen and molybdenum ion implantation in the tribological behavior of AA7075 aluminum alloy. *Wear* **2012**, *276*, 53–60. [CrossRef]
20. Onate, J.I.; Alonso, F.; Garcia, A. Improvement of tribological properties by ion implantation. *Thin Solid Films* **1998**, *317*, 471–476. [CrossRef]
21. Zhao, B.; Li, D.M.; Zeng, F.; Pan, F. Ion beam induced formation of metastable alloy phases in Cu–Mo system during ion beam assisted deposition. *Appl. Surf. Sci.* **2003**, *207*, 334–340. [CrossRef]
22. Suthanthiraraj, S.A.; Premchand, Y.D. Molecular structural analysis of 55mol% CuI-45mol% Ag_2MoO_4, solid electrolyte using XPS and laser Raman techniques. *Ionics* **2004**, *10*, 254–257. [CrossRef]
23. Muratore, C.; Voevodin, A.A.; Hu, J. Tribology of adaptive nanocomposite yttria-stabilized zirconia coatings containing silver and molybdenum from 25 to 700 °C. *Wear* **2006**, *261*, 797–805. [CrossRef]
24. Zhang, Q.-Y.; Zhou, Y.; Wang, L.; Cui, X.-H.; Wang, S.-Q. Investigation on tribo-layers and their function of a titanium alloy during dry sliding. *Tribol. Int.* **2016**, *94*, 541–549. [CrossRef]
25. Kato, H. Effects of supply of fine oxide particles onto rubbing steel surfaces on severe-mild wear transition and oxide film formation. *Tribol. Int.* **2008**, *41*, 735–742. [CrossRef]
26. Gupta, S.; Filimonov, D.; Zaitsev, V.; Palanisamy, T.; El-Raghy, T.; Barsoum, M.W. Study of tribofilms formed during dry sliding of Ta_2AlC/Ag or Cr_2AlC/Ag composites against Ni-based superalloys and Al_2O_3. *Wear* **2009**, *267*, 1490–1500. [CrossRef]
27. Hardell, J.; Hernandez, S.; Mozgovoy, S.; Pelcastre, L.; Courbon, C.; Prakash, B. Effect of oxide layers and near surface transformations on friction and wear during tool steel and boron steel interaction at high temperatures. *Wear* **2015**, *330–331*, 223–229. [CrossRef]
28. Stone, D.; Liu, J.; Singh, D.P.; Muratore, C.; Voevodin, A.A.; Mishra, S. Layered atomic structures of double oxides for low shear strength at high temperatures. *Scr. Mater.* **2010**, *62*, 735–738. [CrossRef]

Article

Enhanced Tribological Properties of Polymer Composite Coating Containing Graphene at Room and Elevated Temperatures

Yanchao Zhang [1], Dongya Zhang [1,2,*], Xian Wei [3], Shanjun Zhong [2] and Jianlei Wang [1]

1 School of Mechanical and Precision Instrument Engineering, Xi'an University of Technology, Xi'an 710048, China; zhangyanchao@xaut.edu.cn (Y.Z.); jlwang@xaut.edu.cn (J.W.)
2 Ningbo Zhongyi Hydraulic Motor Co., Ltd., Ningbo 315200, China; 13506848082@126.com
3 School of Automobile and Traffic Engineering, Panzhihua University, Panzhihua 617000, China; weixian0506@126.com
* Correspondence: dyzhang@xaut.edu.cn

Received: 12 January 2018; Accepted: 26 February 2018; Published: 2 March 2018

Abstract: To improve the tribology properties of the polymer coating under elevated temperature, the epoxy coating was reinforced with nano graphene. The micro-hardness, heat conductivity, and thermo-gravimetric properties of the coating were enhanced as filled graphene. The friction and wear properties of the polymer coating were studied using a pin-on-disc tribo-meter under room and elevated temperatures. The results showed that under room temperature, the friction coefficient and the wear rate of the coating adding 4.0 wt % graphene was 80% and 76% lower than that of the neat epoxy coating, respectively. As the test temperature increased, the friction coefficient of the graphene/polymer coatings decreased at first and then slightly increased. The friction coefficient was at its lowest value under 150 °C and then increased as the temperature rose to 200 °C. By adding 4.0 wt % graphene, the friction coefficient and wear rate of the polymer coating were further reduced, especially at elevated temperatures.

Keywords: polymer coating; graphene; tribological properties; elevated temperature

1. Introduction

In space environments, large frictional heat is difficult to dissipate from sliding surfaces. Additionally, a protective oxide film is difficult to form on contact surfaces during the rubbing process. These issues harm the adhesive and the adhesive wear can lead to failure of moving parts in certain conditions. In addition, the liquid lubricants are easily evaporated under high vacuum and high temperature conditions [1], which made the liquid lubricant hard to maintain on the contact area in order to provide an effective long term lubrication performance. The liquid lubricants are limited under extremely harsh working conditions such as high vacuum or high and/or low temperature [2]. Therefore, the reliability of solid or liquid lubricants to improve the lubrication performance of the moving parts is of great significance for long-life spacecraft. Polymer composite coatings have many advantageous behaviors, such as being able to operate with good oxidation resistance, being able to withstand acid and alkali due to corrosion resistance, and having chemical stability and anti-friction lubricity. As such, polymer composite coatings are often employed in the machinery of bearings, gears, and other moving parts in the space vehicle such as satellite antenna drive system and solar panel equipment.

The major setbacks of polymer composite coatings are usually their high friction coefficient and poor wear resistance. Such poor tribological performance of polymers can be improved by adequately incorporating nano or micro particles and fibers into the polymer matrix. Our previous work [3]

showed the possible improvement of polymer tribological behaviors by appropriately embedding graphite and MoS_2 particles. Jitendra [4] improved the friction and wear behavior of polymer composite by adding talc and graphite powders and showed that the friction coefficient and wear rate of SU-8 composites decreased as comprehensive usage of talc and graphite. Wan [5] utilized Ag nanoparticles to improve tribological properties of polyimide/epoxy resin-polytetrafluoroethylene (denoted as PI/EP-PTFE) coating, and confirmed the friction reduction and the enhancement of wear resistance of PI/EP-PTFE coating by the addition of Ag nanoparticles. A large number of available research studies indicate that the type and size of fillers, as well as the test conditions, had strongly influenced tribological performances of polymer composites [6,7]. However, these fillers usually aggregate with a relatively large size and are poorly dispersed in the polymer matrix, which is unfavorable for enhancing the mechanical properties of polymer coatings [8]. Some polymer coatings also have disadvantages of low hardness and high temperature creep. Due to this, the thermostability and tribological properties are clearly deteriorate in extreme environments.

Graphene is a two-dimensional crystal consisting of carbon atoms that has excellent mechanical, thermal, and tribology properties. Additionally, the nano size with properties of larger surface areas and rich oxygen containing functional groups can enhance the adhesive between the fillers and the chains in the polymer matrix [9], which is beneficial for improving the thermostability and wear resistance of the polymer composite. Meanwhile, graphene can form a self-lubrication transfer film on the contact interfaces during the friction process, which endows the graphene/polymer composite with a low and stable friction coefficient and wear rate [10–12].

When Liu [13] added graphene into the polyimide, the thermal stability and the hardness of the polyimide were significantly improved. With the addition of graphene content, the adhesive strength of the transfer film was increased. Compared with the neat polyimide, when the graphene content reached 3.0 wt %, the friction coefficient and wear rate decreased 21.3% and 26.3%, respectively. Ren [14] synthesized functionalized graphene and employed it as filler to improve the anti-wear property and load-carrying capacity of fabric/phenolic composites. It was found that the 2.0 wt % graphene filled fabric/phenolic composite exhibited excellent tribological properties. Lahiri [15] reinforced the tribological behavior of ultrahigh molecular weight polyethylene by coupled with graphene platelet. The wear resistance was enhanced more than four times as increasing graphene content from 0.1 to 1.0 wt %. Recently, Masood [16] added graphene and PTFE to improve the tribological response of nylon-based composites, and an optimal graphene content (0.5% in weight) could synergistically improve the friction coefficient and wear rate of Nylon 66. Numerous studies showed that graphene had excellent lubrication and anti-wear performance, and graphene was widely used as filler to improve the mechanical property and tribological performance of the polymer composites [17–19]. The enhancement on the wear property of the composite was mainly due to the self-lubrication of graphene and the easily-formed transfer film on the counterpart surface. However, the tribological mechanisms of graphene/polymer coating are not well understood and thus require further investigation. Thermo-stability and the tribological properties are important properties of polymers. The full exploration of lubrication behaviors of polymer coating and the understanding of their tribological applications under high temperatures are very meaningful and significant.

Epoxy resin is widely used for composite coating, and the polymer coating with good adhesion performance can be conveniently prepared on the surface by a spray process. However, under a poor heat dissipation condition, huge frictional heat is hard to dissipate in time, which influences the lifetime of the polymer coating. In this work, in order to enhance the high temperature tribological performance of polymer coating, graphene was prepared and incorporated into epoxy resin matrix. The main objective of this study was to investigate the influences of graphene content and test temperature on tribological properties of the polymer coating, as well as to discuss their corresponding lubrication and wear mechanisms.

2. Materials and Methods

2.1. Preparation of Polymer Composite Coating

The detailed process of synthesis graphene was as follows. Graphene oxide was first prepared by the oxidation–deoxidization method according to a modified Hummer's method [20]. To fabricate graphene oxide:(i) 120 mL sulfuric acid with 98% concentration was added into the beaker and cooled in the ice bath, then 5 g flake graphite (300 mesh) and 2.5 g $NaNO_3$ were slowly added in the sulfuric acid solution, and then 15 g $KMnO_4$ was gradually joined in the solution, the magnetic agitation was applied during the reaction for 1.5 h; (ii) the beaker was heated to 35 °C in a warm bath, meanwhile 200 ml distilled water was slowly add into the mixture and heated 1 h; (iii) the mixture was heated to 95 °C, the reaction was lasted for 30 min to get tan precipitation, which was the graphene oxide (GO). Finally, graphene oxide was ultrasonic cleaning in deionized water, and dried in vacuum oven.

For graphene: (i) 6.5 g GO (graphene oxide) was added into 1000 mL deionized water and later ultrasonic dispersed for 2 h to obtain a stable graphene oxide solution; (ii) pre-measured 60 mL ammonia (acted as a complexing agent) and 65 mL 85% hydrazine hydrate (acted as a reducing agent) were added dropwise to the graphene oxide solution, respectively. The obtained solution was heated at 80 °C for 2 h, and then the reaction system was cooled down to room temperature after the reaction completed; (iii) the obtained graphene was washed five times with deionized water by centrifugation at 3000 rpm over 15 min, then the solution was frozen in the refrigerator for 72 h until the graphene was freeze dried.

GCr15 steel with a diameter of 30 mm and a thickness of 5 mm was applied as a substrate specimen, and the disc specimen had hardness of HRC 60–63. The specimens were polished with 100 grain size sand paper to increase the surface roughness. The disc specimens to be coated with polymer coating were cleaned in an ultrasonic bath with acetone for 10 min successively three times so as to ensure the proper removal of residual pollutants. Then the specimens were dried at 100 °C before spraying the coating. The polymer coating was achieved by spraying epoxy matrix filled with graphene, the epoxy matrix was high purity E51 epoxy resin with an epoxide equivalent weight of 210–250 g/eq. 650 type low molecular polyamine (with amine value of 180–220 mg KOH/g) was employed as the curing agent for the epoxy monomer.

The graphene (content increases from 0.0% to 4.0% in weight) were slowly added into the E51 epoxy resin at 70 °C and uniformly stirred. Then the curing agent was introduced into the mixture at 60 °C, and the solution was continuously mixed for 10 min and diluted with ethyl alcohol. The prepared graphene–epoxy solution was then sprayed on the GCr15 substrate, and it was followed by curing at room temperature for 24 h. Finally, the composite coating with a thickness 30 μm and surface roughness R_a of 0.5 μm was prepared.

2.2. Characterization

Micro-hardness of the polymer coating was performed by using a TMVS-1 hardness tester (TIMES Group, Beijing, China). For each specimen, five measurements were conducted under load of 50 gf and lasting 15 s. Thermal gravimetric analysis (TG) of the polymer coatings filled with different contents of graphene were evaluated using a TG apparatus (Netzsch Sta 409 PC/PG, Netzsch, Selb, Germany). 3.5 mg specimen was heated from 25 to 750 °C at a heating rate of 10 °C/min in nitrogen atmosphere. Thermal conductivity of the polymer composite was carried out using a LFA 447 Nanoflash (Netzsch, Selb, Germany), according to ASTM E1461 standard [21]. The sample was prepared with a size of 10 mm diameter and 1 mm thickness. Before each experiment, a thin graphite layer was coated on two sides to increase emission/absorption behavior. The test was performed at a room temperature of 25 °C, the thermal diffusivity values (cm^2/s) of the composites were recorded.

To evaluate the tribological properties of polymer coating from room temperature to 200 °C, the disc specimen was heated and carried out on UMT-2 tribometer (CETR Corporation Ltd., Campbell, CA, USA). Linearly reciprocating ball-on-disk sliding tests were performed according to ASTM G133-05 [22], the lower specimen was the polymer coating and the upper specimen was a GCr15 ball with diameter of 9.5 mm and hardness of 62 HRC. Before the tests, the ball specimens were ultrasonically cleaned in deionized water for 5 min and the disc specimens were cleaned with alcohol wipes, and then dried in hot air. The test conditions were as follows: applied load of 4 N, reciprocating sliding frequency of 6 Hz, with a liner stroke of 6 mm, and sliding time of 60 min. The friction tests were carried out under 25, 100, 150, and 200 °C, respectively, and each test was repeated three times. After the friction tests, the wear scars were observed by a scanning electron microscope (SEM, JSM-6460, JEOL, Tokyo, Japan).

Cross-sectional profile of the wear scar was measured using a surface profilometer (TR3000, TIMES Group, Beijing, China), and the wear rate was calculated as the ratio of its wear volume and the corresponding sliding distance and applied load. From the profile curves, this allowed estimation of the sectional area of the wear trace and the average wear width. The average wear width was used to calculate the average worn volume V, and the corresponding wear rate k was calculated in the following equation [23].

$$k = \frac{V}{P \times S} = \frac{\pi D[(\arcsin\frac{L}{2r})r^2 - \frac{L\sqrt{4r^2-L^2}}{4}]}{P \times S} \quad (1)$$

where k is the wear rate (mm^3/N·m), V is the wear volume (mm^3), r is the ball radius (3 mm), D is the diameter of wear track (10 mm), L is the width of wear scar (mm), S is the sliding distance (m), and P is the applied load (N).

3. Results and Discussion

3.1. Characterization of the Composite Coating

The surface morphology of graphene is characterized using a JSM-6460 scanning electron microscope (SEM), as shown in Figure 1a. It shows that the lateral dimensions of the graphene are layered with a nano-scale thickness. However, some graphene particles agglomerate with each other, and form a corrugated structure. The fracture surface of the composite coating filled with graphene is shown in Figure 1b. This shows that the graphene is tightly combined in the epoxy matrix without loosening even after the fracture. The graphene has a larger specific surface area, which assures strong anchoring and increases the inter-facial contact between the graphene and polymer. It indicates that the presence of well-dispersed graphene in the polymer composite can lead to desirable bearing properties and improve the tribological performance.

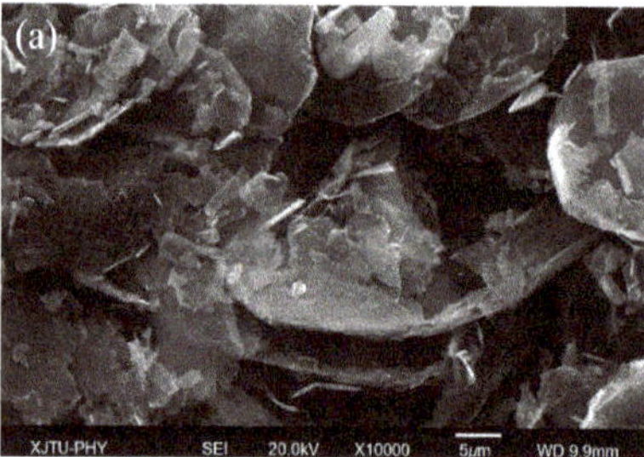

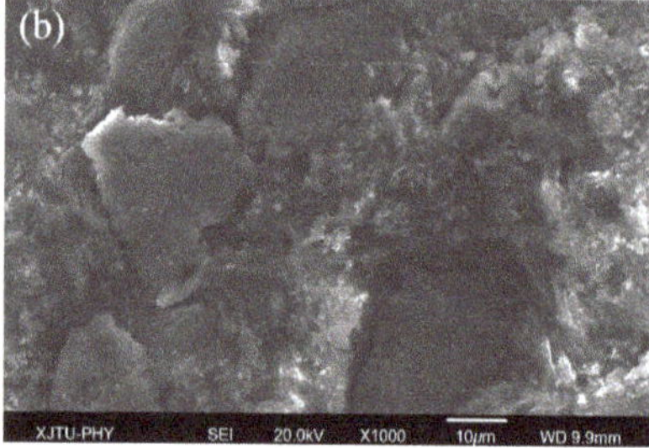

Figure 1. SEM images of (**a**) graphene and (**b**) the fracture surface of polymer coating (4.0 wt % graphene).

Figure 2 shows the micro-hardness of polymer coating depending on various graphene contents. The hardness is gradually increased when the content of graphene increases from 0.0 to 4.0 wt %. When the graphene content is higher than 2.0 wt %, the micro-hardness increases slightly with the increase of graphene content. The lowest hardness of the neat epoxy coating is 20.4 Hv, while the hardness is 37 Hv as the graphene content increases to 4.0 wt %. The increase in graphene content caused hardness to rise, this result is consistent with previous works [24]. The addition of a coupling agent can effectively enhance the adhesive strength between epoxy matrix and fillers [25]. Homogeneous dispersion of fillers improves the hardness of the polymer coating. The improvement in hardness is mainly due to the formation of a three-dimensional network by the uniform and staggered distribution of graphene in a polymer matrix. When increasing graphene content in the polymer matrix, it is beneficial to enhance the wear resistance of the composite.

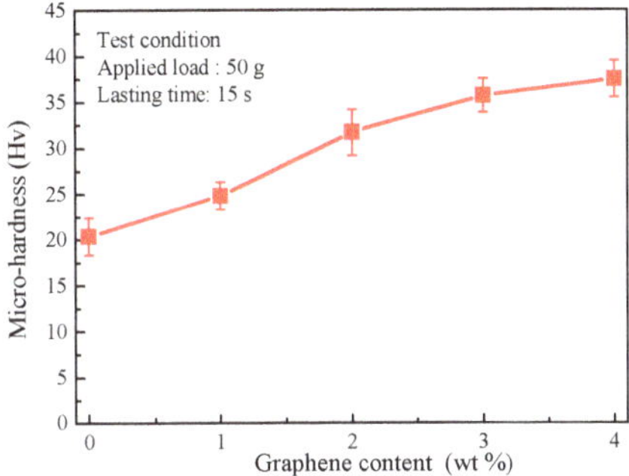

Figure 2. Effect of graphene content on the micro-hardness of the polymer composite coatings.

The thermal conductivities of the polymer composite measured by the laser flash method are plotted in Figure 3. It shows that the thermally conductive coefficient of the polymer composite is significantly enhanced to 2.36 W/m·K by adding 4.0 wt % graphene, which is 12 times higher than that of neat epoxy. The theoretical thermal conductivity of graphene is reported to be as high as 5000 W/m·K [26]. Therefore, it is reasonable to assume graphene is suitable for fabricating the epoxy nanocomposite with high thermal conductivity. It also indicates that there is an obvious increment for the thermal conductivity as an increase of graphene content. Because heat propagation in the polymer is mainly due to acoustic phonons, a uniform network in the polymer matrix may result in an increase in thermal conductivity in the composites [27]. With the increasing graphene content, the graphene particles connect to each other and the thermally conductive network is easily formed. For a high content of graphene, polymer composite possesses better interfacial compatibility, which favors phonon transport [28], and thus increases thermal conductivity. Similar conclusions in thermal conductivity of epoxy matrix was previously reported by [27,29].

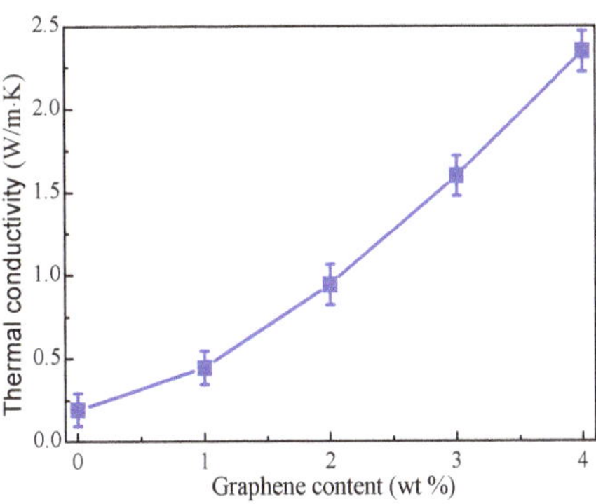

Figure 3. Effect of graphene content on the thermal conductivity of the polymer composite coatings.

Thermal properties of polymer coating with various graphene content by thermal gravimetric (TG) analysis is shown in Figure 4. When the specimen is heated, there are two main stages of mass loss of polymer composite during the thermal gravimetric analysis. In the temperature range of 50–300 °C, all the TG curves of the specimens present almost the same change trend, the weight is slowly decreased with a rise in temperature, which is caused by the volatilization of trace amounts of water and organic matter in the matrix. When the temperature is 300–400 °C, the weight loss increases rapidly as the temperature rises. Decomposition tendencies with temperature are similar, this is due to thermal decomposition of main chain of epoxy. In the range of 400–700 °C, the decomposition speeds are slow, and the weight loss is significantly different. The decomposition rate of epoxy resin with 4.0 wt % graphene is slow, and the residue yields of degradation are highest at the temperature of 700 °C, which shows that the heat resistant properties of epoxy resin with 4.0 wt % graphene are superior. With the increase of graphene content, the heat resistant performance of polymer coating gradually improves. It indicates that the graphene is effective for enhancing the thermal stability of epoxy resin. The enhancement of thermal stability can be explained in terms of the dispersion of graphene and interfacial interaction with the epoxy matrix [30]. As such, the addition of graphene is helpful for improving the thermal stability of the polymer matrix in the high temperature stage.

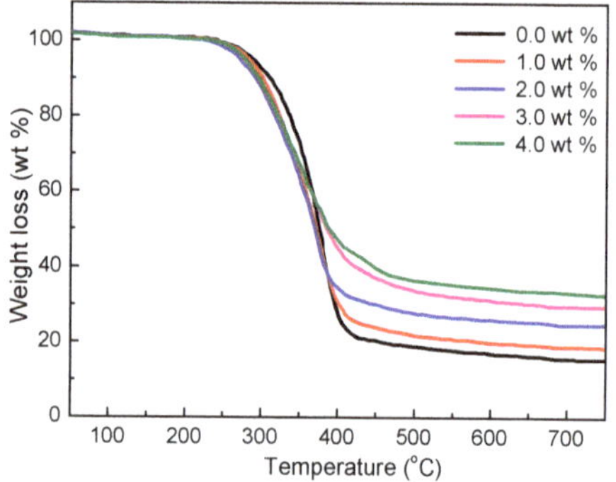

Figure 4. TG curves of polymer composite coatings filled with various grapheme contents.

3.2. *Effect of Graphene Content on the Tribological Properties*

The effect of the graphene on the tribological properties of the reinforced polymer coating was investigated by determining the value of the friction coefficient and wear rate with various graphene contents (1.0–4.0 wt %, respectively).

Figure 5a shows the friction coefficient curves of the polymer coatings with various content of graphene at room temperature. The friction coefficient of neat epoxy coating is 0.41 at the start-up stage, and then quickly increases to 0.50 after running for 100 s. The increment rises slowly when the sliding continues, the friction coefficient rises to 0.60 at the end of the test. The friction coefficients of composite coatings are significantly reduced and more stable when adding graphene. For example, in the case of adding 1.0 wt % graphene, the friction coefficient of coating reduces to 0.25, which is 50% lower than that of neat epoxy coating. It also confirms that the friction coefficient gradually decreases as the content of graphene increases. Furthermore, when the graphene content is 4.0 wt %, the friction coefficient (with the lowest value of 0.11) is significantly lower than that of the other coatings. Figure 5b illustrates that the wear rate of the polymer coating varies with the content of graphene at room temperature. The wear rate of the neat epoxy coating is as high as 6.54×10^{-6} mm^3/N·m. When the graphene content is 1.0 wt %, its wear rate is reduced by 59% when compared with neat epoxy coating, this indicates an effective improvement for wear resistance of the polymer composite at a relatively low content. In addition, it can be seen that the wear rate gradually decreases as the graphene content increases.

The significant decrease in the friction coefficient and wear rate for polymer composite coating can be associated with the self-lubrication of graphene. When the embedded graphene in the polymer matrix is extruded and it forms solid stable transfer films on the relative sliding surfaces, which can prevent the sliding occurring between the rough surface of composite coatings and steel counterpart, it endows a self-lubricating characteristic of the polymer composite [31,32].

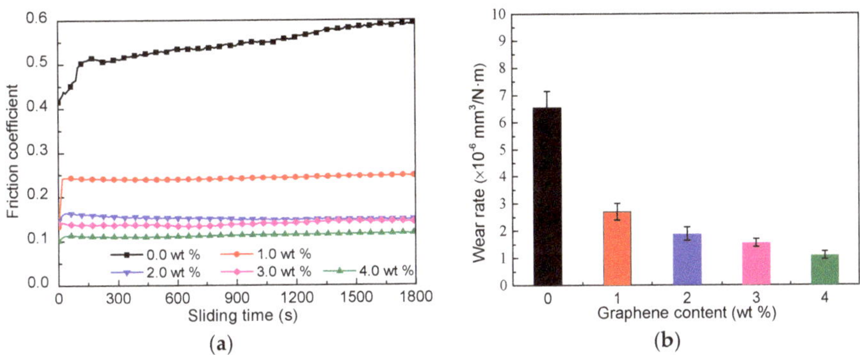

Figure 5. Tribology properties of polymer composite coatings: (**a**) friction coefficient and (**b**) wear rate.

3.3. *Effect of Elevated Temperature on the Tribological Properties*

Figure 6 shows the average friction coefficient of the polymer coating with various graphene contents slid against the GCr15 steel ball with temperatures varying from 25 to 200 °C.

It can be seen that the neat epoxy coating has the highest friction coefficient, and the friction coefficient first reduces and then increases as the test temperature rises. The friction coefficient is lowest (0.44) at 150 °C and then increases to 0.48 as the temperature rises to 200 °C. It is also found that the friction coefficient of the coating containing 1.0 wt % graphene is stable with a rise in test temperature, the average friction coefficient is in the range of 0.25–0.28, which is distinctly lower than the neat epoxy coating. In addition, the graphene content further increases to 3.0 wt %, the friction coefficient declines while the reduction rate slows down as the test temperature increases, the lowest friction coefficient is 0.07 at the test temperature of 150 °C. Furthermore, when the graphene content increases to 4.0 wt %, the friction coefficient is lower than the coating contents of 3.0 wt % graphene under the same test conditions. Generally, high graphene content benefits the high temperature tribological performance of polymer coating, whereas higher test temperature leads to a higher friction coefficient.

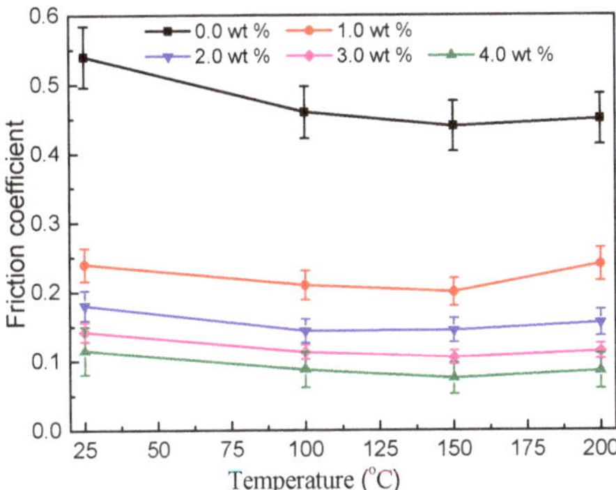

Figure 6. Influence of temperature on the friction coefficient of the polymer composite coatings.

The wear rate of polymer coating with various graphene contents under a test temperature range from 25 to 200 °C, which is shown in Figure 7. It indicates that polymer coating without graphene has the highest wear rate, and the wear rate reduces and then increases with temperature increases, the minimum wear rate is 5.62×10^{-6} mm^3/N·m at 100 °C, and the wear rate reaches the highest value (9.75×10^{-6} mm^3/N·m) at 200 °C. When the graphene content increases to 1.0 wt %, the wear rate is significantly lower than neat epoxy coating. Particularly, the wear rate of the coating with a graphene content of 4.0 wt % is lowest in the four grapheme-containing polymer coatings. In the given contents, the wear resistance of the polymer coating is better under higher graphene content. Thus, the graphene improves the wear resistance of polymer coating under higher temperature. When graphene is intercalated in epoxy resin, and the organic-inorganic hybrid structure is formed, this prevents the heat flow to the epoxy resin and reduces the thermal decomposition of composite matrix, so the wear resistance is significantly enhanced at high temperature stages. Prior study also found that adding graphene in the epoxy matrix could enhance the bearing capacity and the fatigue strength of the polymer [15].

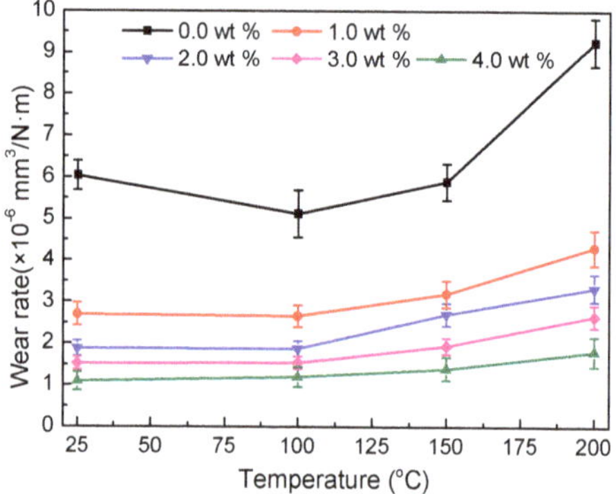

Figure 7. Wear rate of the polymer composite coatings under various test temperatures.

Neat epoxy polymer is viscoelastic material, the friction mainly depends on the adhesion of the epoxy and steel on the contact area. The relaxation of the branched chain of epoxy resin [33] is occurred under proper heat, and shear slip of the epoxy takes place under the rubbing process along the sliding direction and decreases friction. Therefore, friction coefficient of neat epoxy under 150 °C is lower than that under 25 °C. However, as the temperature continues to rise, the hardness declines and the real contact area also increases, which causes severe wear and results in further increase of the friction coefficient of neat polymer.

However, the friction coefficient of the polymer composite coating with the addition of graphene is slowly decreased or increased as the increase of test temperatures. This can be attributed to the embedded particles are free to roll, since the viscous polymer cannot hold them under higher temperature. The rolling particles could significantly reduce friction and temperature in the contact area [34,35]. Furthermore, the graphene also forms a transfer film on the counterpart surfaces. As a result, both the frictional coefficient and the wear rate of the polymer composite coating are effectively reduced.

3.4. Analysis of Wear Morphology

Figure 8 shows SEM micrographs of the worn surface of epoxy coating with various graphene contents. For the neat epoxy coating, a large amount of micro-cracking and material breaking off are observed on the wear track, which indicates a typical fatigue wear type of the neat epoxy coating under room temperature (Figure 8a). As graphene is added into the epoxy matrix, the damage of the wear decreases because the contact stress is supported by the graphene. When the content of graphene is 1.0 wt %, the wear scratch on the surface gets smoother and shows no surface defects, and only a smooth and very shallow furrow. When the content of graphene is 3.0 and 4.0 wt %, the worn surfaces of polymer coating (Figure 8d,e) can be characterized similarly, which is a wear feature of the surface that exhibited plastic deformation.

When graphene content is increased, the wear damage of the polymer coating decreases, the reason is that a high filing ratio of graphene in the epoxy matrix can form a well-dispersed graphene-epoxy structure, which facilitates good load transfer to the matrix network, resulting in improvement of tribologilcal properties of the epoxy coating.

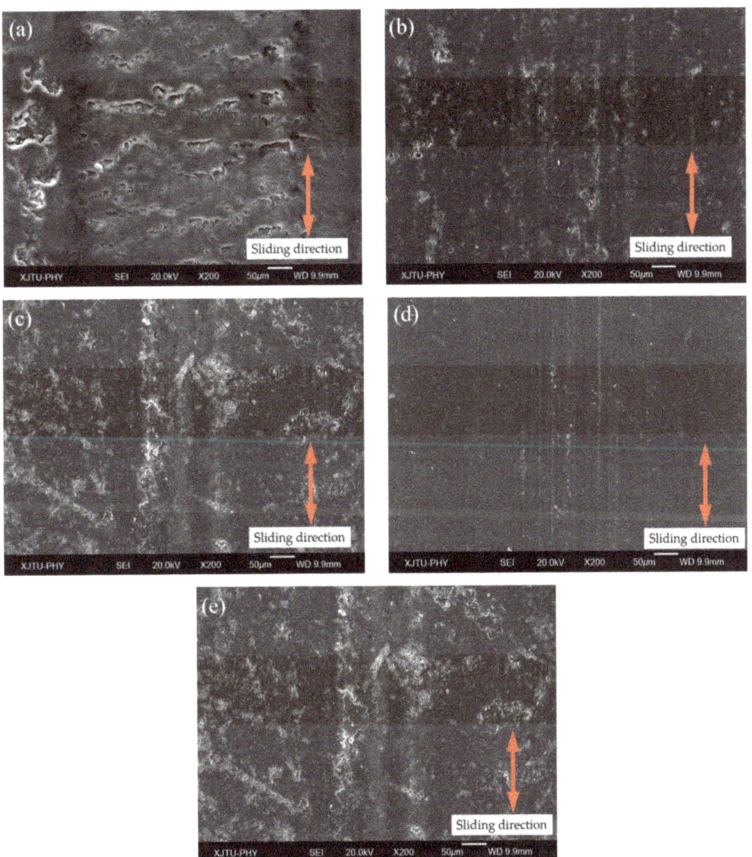

Figure 8. SEM images of worn surfaces of epoxy coating with the graphene content of (**a**) 0.0 wt %, (**b**) 1.0 wt %, (**c**) 2.0 wt %, (**d**) 3.0 wt %, and (**e**) 4.0 wt %, respectively, at room temperature.

The micrographs of their wear scars slid obtained under test temperature of 150 °C are compared to better understand the possible wear mechanisms of epoxy coating under high temperature. The SEM images of the worn surfaces are shown in Figure 9. The worn surface of the neat epoxy is composed of a loose debris layer, and a large number of micro cracks and peeling, which indicates that the neat epoxy coating experienced intense fatigue failure under high thermal stress (Figure 9a). When the graphene content increases to 1.0 wt %, it can be seen that wear debris are sheared by a micro convex body, and then roller compacted to form a transfer film on the worn surface, which can support the contact press. This also indicates that the continuous transfer film is hard to form under a low grapheme content. As shown in Figure 9c, for the graphene content of 3.0 wt %, the wear becomes relatively smooth and continuous transfer film is formed on the surface. The detection facilitates to deduce that in the sliding process, graphene can form a transfer film on the friction pair, and graphene in the polymer matrix can improve the thermal conduction performance, as well as avoid the initial failure of coating. Adding an appropriate amount of graphene can improve the wear resistance of polymer coating mainly because the graphene enhances the bearing strength of the coating. Therefore, the coating containing graphene can withstand a larger shear force. In addition, the micrographs of worn surface phenomena decrease with increasing graphene content.

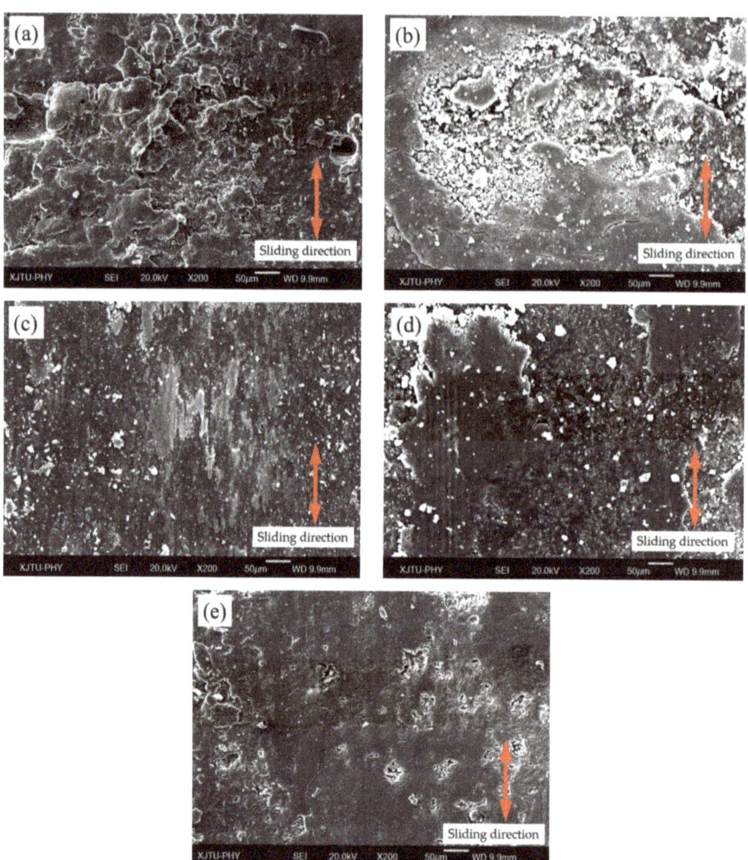

Figure 9. SEM images of worn surfaces of epoxy coating with the graphene content of (**a**) 0.0 wt %, (**b**) 1.0 wt %, (**c**) 2.0 wt %, (**d**) 3.0 wt %, and (**e**) 4.0 wt % respectively under 150 °C.

4. Conclusions

Graphene reinforced polymer coating was successfully deposited on a GCr15 steel surface by a spraying process. Investigation of micro-hardness, thermal characteristics, and high temperature tribological properties of the epoxy coating were conducted, and the effects of graphene on the mentioned performances had been discussed. The following conclusions were reached:

- When filling with various content of graphene, the micro-hardness, heat resistance, and thermal conductivity properties of epoxy resin coating are significantly enhanced.

- Under the room temperature condition, epoxy coating containing graphene has excellent tribological properties, the friction coefficient and wear rate of the neat epoxy coating is 0.55 and 6.54×10^{-6} $mm^3/N{\cdot}m$, the friction coefficient and wear rate is 0.11 and 1.52×10^{-6} $mm^3/N{\cdot}m$ for the composite coating contains 4.0 wt % graphene. With the increase of graphene content, the friction coefficient and wear rate of the composite coating are gradually reduced.
- Under high temperature conditions, graphene-enhanced composite coating shows better friction reduction and wear resistance than neat epoxy coating, and the values are reduced as the increase of graphene content. Meanwhile, the friction coefficient and the wear rate of the composite coatings containing graphene slightly decrease or increase with the increase of test temperature. Such phenomena are mainly caused by the formation of a transfer film on the surface which suppresses the huge heat and contact pressure.

Acknowledgments: This work was supported by the National Natural Science Foundation of China (No. 51605381), China Postdoctoral Science Foundation (2017M613171), Shaanxi Provincial Education Department (No. 16JK1540), and Natural Science Fund Plan of Shaanxi Province (No. 2017JQ5101).

Author Contributions: Dongya Zhang conceived and designed the experiments; Xian Wei performed the experiments for fabricating composite coatings; Shanjun Zhong analyzed the tribological experiment results; Yanchao Zhang and Jianlei Wang contributed analysis of the thermal performance of the coating; Dongya Zhang wrote the paper.

Conflicts of Interest: The authors declare no conflict of interest.

References

1. Ke, H.; Huang, W.; Wang, X. Insights into the effect of thermocapillary migration of droplet on lubrication. *Proc. Inst. Mech. Eng. Part J-J. Eng. Tribol.* **2016**, *230*, 583–590. [CrossRef]
2. Donnet, C.; Erdemir, A. Solid lubricant coatings: Recent developments and future trends. *Tribol. Lett.* **2004**, *17*, 389–397. [CrossRef]
3. Zhang, D.-Y.; Zhang, P.-B.; Lin, P.; Dong, G.-N.; Zeng, Q.-F. Tribological properties of self-lubricating polymer-steel laminated composites. *Tribol. Trans.* **2013**, *56*, 908–918. [CrossRef]
4. Katiyar, J.K.; Sinha, S.K.; Kumar, A. Friction and wear durability study of epoxy-based polymer (SU-8) composite coatings with talc and graphite as fillers. *Wear* **2016**, *362–363*, 199–208. [CrossRef]
5. Wan, H.; Jia, Y.; Ye, Y.; Xu, H.; Cui, H.; Chen, L.; Zhou, H.; Chen, J. Tribological behavior of polyimide/epoxy resin-polytetrafluoroethylene bonded solid lubricant coatings filled with in situ-synthesized silver nanoparticles. *Prog. Org. Coat.* **2017**, *106*, 111–118. [CrossRef]
6. Turssi, C.P.; Purquerio, B.M.; Serra, M.C. Wear of dental resin composites: Insights into underlying processes and assessment methods—A review. *J. Biomed. Mater. Res. B* **2003**, *65*, 280–285. [CrossRef] [PubMed]
7. Bijwe, J.; Indumathi, J.; Ghosh, A.K. On the abrasive wear behaviour of fabric-reinforced polyetherimide composites. *Wear* **2002**, *253*, 768–777. [CrossRef]
8. Bastani, D.; Esmaeili, N.; Asadollahi, M. Polymeric mixed matrix membranes containing zeolites as a filler for gas separation applications: A review. *J. Ind. Eng. Chem.* **2013**, *19*, 375–393. [CrossRef]
9. Shen, X.-J.; Pei, X.-Q.; Fu, S.-Y.; Friedrich, K. Significantly modified tribological performance of epoxy nanocomposites at very low graphene oxide content. *Polymer* **2013**, *54*, 1234–1242. [CrossRef]
10. Lin, J.; Wang, L.; Chen, G. Modification of graphene platelets and their tribological properties as a lubricant additive. *Tribol. Lett.* **2011**, *41*, 209–215. [CrossRef]
11. Kandanur, S.S.; Rafiee, M.A.; Yavari, F.; Schrameyer, M.; Yu, Z.Z.; Blanchet, T.A.; Koratkar, N. Suppression of wear in graphene polymer composites. *Carbon* **2012**, *50*, 3178–3183. [CrossRef]
12. Liu, D.; Zhao, W.S.; Liu, Q.; Cen, Q.; Xue, Q. Comparative tribological and corrosion resistance properties of epoxy composite coatings reinforced with functionalized fullerene C60 and graphene. *Surf. Coat. Technol.* **2016**, *286*, 354–364. [CrossRef]
13. Liu, H.; Li, Y.; Wang, T.; Wang, Q. In situ synthesis and thermal, tribological properties of thermosetting polyimide/graphene oxide nanocomposites. *J. Mater. Sci.* **2012**, *47*, 1867–1874. [CrossRef]

14. Ren, G.; Zhang, Z.; Zhu, X.; Ge, B.; Guo, F.; Men, X.; Liu, W. Influence of functional graphene as filler on the tribological behaviors of Nomex fabric/phenolic composite. *Compos. Part A-Appl. Sci. Manuf.* **2013**, *49*, 157–164. [CrossRef]
15. Lahiri, D.; Hec, F.; Thiesse, M.; Durygin, A.; Zhang, C.; Agarwal, A. Nanotribological behavior of graphene nanoplatelet reinforced ultrahigh molecular weight polyethylene composites. *Tribol. Int.* **2014**, *70*, 165–169. [CrossRef]
16. Muhammad, T.M.; Evie, L.P.; José, A.H.G.; Ilker, S.B.; Athanassi, A.; Luca, C. Graphene and polytetrafluoroethylene synergistically improve the tribological properties and adhesion of nylon 66 coatings. *Carbon* **2017**, *123*, 26–33.
17. Shen, X.-J.; Pei, X.-Q.; Liu, Y.; Fu, S.-Y. Tribological performance of carbon nanotube–graphene oxide hybrid/epoxy composites. *Compos. Part B-Eng.* **2014**, *57*, 120–125. [CrossRef]
18. Jiang, T.; Kuila, T.; Kim, N.H.; Ku, B.-C.; Lee, J.H. Enhanced mechanical properties of silanized silica nanoparticle attached graphene oxide/epoxy composites. *Compos. Sci. Technol.* **2013**, *79*, 115–125. [CrossRef]
19. Golchin, A.; Wikner, A.; Emami, N. An investigation into tribological behaviour of multi-walled carbon nanotube/graphene oxide reinforced UHMWPE in water lubricated contacts. *Tribol. Int.* **2016**, *95*, 156–161. [CrossRef]
20. Marcano, D.C.; Kosynkin, D.V.; Berlin, J.M.; Sinitskii, A.; Sun, Z.; Slesarev, A.; Alemany, L.B.; Lu, W.; Tour, J.M. Improved synthesis of graphene oxide. *ACS Nano* **2010**, *4*, 4806–4814. [CrossRef] [PubMed]
21. *ASTM E1461-01 Standard Test Method for Thermal Diffusivity by the Flash Method*; ASTM International: West Conshohocken, CA, USA, 2001.
22. *ASTM G133-05 Standard Test Method for Linearly Reciprocating Ball-on-flat Sliding Wear*; ASTM International: West Conshohocken, CA, USA, 2005.
23. Zhang, D.; Dong, G.; Chen, Y.; Zeng, Q. Electrophoretic deposition of PTFE particles on porous anodic aluminum oxide film and its tribological properties. *Appl. Surf. Sci.* **2014**, *290*, 466–474. [CrossRef]
24. Tai, Z.; Chen, Y.; An, Y.; Yan, X.; Xue, Q. Tribological behavior of UHMWPE reinforced with graphene oxide nanosheets. *Tribol. Lett.* **2012**, *46*, 55–63. [CrossRef]
25. Naves, L.Z.; Santana, F.R.; Castro, C.G.; Valdivia, A.D.C.M.; da Mota, A.S.; Estrela, C.; Sobrinho, L.C.; Soares, C.J. Surface treatment of glass fiber and carbon fiber posts: SEM characterization. *Microsc. Res. Tech.* **2011**, *74*, 1088–1092. [CrossRef] [PubMed]
26. Liu, L.H.; Yan, M. Functionalization of pristine graphene with perfluorophenyl azides. *J. Mater. Chem.* **2012**, *21*, 3273–3276. [CrossRef]
27. Chatterjee, S.; Wang, J.W.; Kuo, W.S.; Tai, N.H.; Salzmann, C.; Li, W.L.; Hollertz, R.; Nüesch, F.A.; Chu, B.T.T. Mechanical reinforcement and thermal conductivity in expanded graphene nanoplatelets reinforced epoxy composites. *Chem. Phys. Lett.* **2012**, *531*, 6–10. [CrossRef]
28. Lin, Z.; Mcnamara, A.; Liu, Y.; Moon, K.S.; Wong, C.P. Exfoliated hexagonal boron nitride-based polymer nanocomposite with enhanced thermal conductivity for electronic encapsulation. *Compos. Sci. Technol.* **2014**, *90*, 123–128. [CrossRef]
29. Wang, F.; Drzal, L.T.; Qin, Y.; Huang, Z. Mechanical properties and thermal conductivity of graphene nanoplatelet/epoxy composites. *J. Mater. Sci.* **2015**, *50*, 1082–1093. [CrossRef]
30. Ganguli, S.; Roy, A.K.; Anderson, D.P. Improved thermal conductivity for chemically functionalized exfoliated graphite/epoxy composites. *Carbon* **2008**, *46*, 806–817. [CrossRef]
31. Pan, B.; Zhang, S.; Li, W.; Zhao, J.; Liu, J.; Zhang, Y.; Zhang, Y. Tribological and mechanical investigation of MC nylon reinforced by modified graphene oxide. *Wear* **2012**, *294–295*, 395–401. [CrossRef]
32. Min, C.; Nie, P.; Song, H.J.; Zhang, Z.; Zhao, K. Study of tribological properties of polyimide/graphene oxide nanocomposite films under seawater-lubricated condition. *Tribol. Int.* **2014**, *80*, 131–140. [CrossRef]
33. Feng, J.; Guo, Z. Temperature-frequency-dependent mechanical properties model of epoxy resin and its composites. *Compos. Part B-Eng.* **2016**, *85*, 161–169. [CrossRef]
34. Li, C.; Zhong, Z.; Lin, Y.; Friedrich, K. Tribological properties of high temperature resistant polymer composites with fine particles. *Tribol. Int.* **2007**, *40*, 1170–1178.
35. Rapoport, L.; Bilik, Y. Hollow nanoparticles of WS_2 as potential solid-state lubricants. *Nature* **1997**, *387*, 791–793. [CrossRef]

Article

On the Importance of Combined Scratch/Acoustic Emission Test Evaluation: SiC and SiCN Thin Films Case Study

Jan Tomastik [1,*], Radim Ctvrtlik [2,*], Martin Drab [3] and Jan Manak [4]

1. Regional Centre of Advanced Technologies and Materials, Joint Laboratory of Optics of Palacky University and Institute of Physics of Academy of Sciences of the Czech Republic, Faculty of Science, Palacky University, 17. listopadu 12, 771 46 Olomouc, Czech Republic
2. Institute of Physics of the Academy of Sciences of the Czech Republic, Joint Laboratory of Optics of Palacky University and Institute of Physics AS CR, 17. listopadu 50a, 772 07 Olomouc, Czech Republic
3. ZD Rpety—Dakel, Ohrobecká 408/3, 142 00 Prague, Czech Republic; martin.drab@fjfi.cvut.cz
4. Department of Material Analysis, Institute of Physics CAS, Na Slovance 2, 182 21 Prague, Czech Republic; manak@fzu.cz

* Correspondence: tomastik@jointlab.upol.cz (J.T.); ctvrtlik@fzu.cz (R.C.); Tel.: +420-585-631-573 (J.T.)

Received: 9 April 2018; Accepted: 21 May 2018; Published: 22 May 2018

Abstract: The scratch test, as probably the most widespread technique for assessment of the adhesive/cohesive properties of a film–substrate system, fully depends on reliable evaluation based on assessment of critical loads for systems' failures. Traditionally used evaluation methods (depth change record and visual observation) may sometimes give misleading conclusions about the failure dynamics, especially in the case of opaque films. Therefore, there is a need for another independent evaluation technique with the potential to complete the existing approaches. The nondestructive method of acoustic emission, which detects the elastic waves emitted during film cracking and delamination, can be regarded as a convenient candidate for such a role even at nano/micro scale. The strength of the combination of microscopic observation of the residual groove and depth change record with the acoustic emission detection system proved to be a robust and reliable approach in analyzing adhesion/cohesion properties of thin films. The dynamics of the gradual damage taking place during the nano/micro scratch test revealed by the combined approach is presented for SiC and SiCN thin films. Comparison of critical load values clearly reflects the higher ability of the AE approach in detecting the initial material failure compared to the visual observation.

Keywords: scratch test; acoustic emission; thin films; silicon carbide

---

1. Introduction

Rising demands for modern materials with application-optimized physical properties are the main driving force in contemporary material engineering. The use of surface films and coatings is a logical and successful solution to this challenge as materials and components interact with their surroundings through their surface. Nowadays, various types of physical vapor deposition (PVD), chemical vapor deposition (CVD), and sol-gel techniques are used for production of films in nearly all branches of industry [1–9]. Regardless of the primary functional property of the films predetermined by their application, their mechanical properties and stability are also essential for their long-term service. Mainly, adhesion of the films to the substrate is the primal parameter governing the performance of the system.

The definition given by the American Society for Testing and Materials (ASTM) says that "adhesion is the state in which two surfaces are held together by interfacial forces which may consist of valence forces or interlocking forces or both" [10]. This "basic or also theoretical adhesion" depends on the electrostatic, chemical, and/or van der Waals forces, which act on the whole contact surface between film and substrate. However, the measured "experimental adhesion" exhibits significantly smaller values due to structural voids and material peculiarities (internal stresses, porosity, homogeneity, roughness of contact area, etc.) and also due to the effects associated with the measurement methods (fixing of the sample, tangential forces, experimental errors, etc.) as well as the rigidity of the testing instrument itself [11,12]. Various experimental techniques have been introduced, aiming for not only the qualitative differentiation between the measured samples but also for the quantification of adhesion forces. In addition to the non-mechanical methods (thermal, nucleation, x-ray diffraction, etc.) [13–15], the more direct mechanical methods (scotch tape, bend test, laser spallation, direct pull-off, etc.) [12,15–18] have been used with varying success. Nevertheless, their mutual comparison is questionable and the level of proximity to the basic adhesion varies. It is mainly the difficulty of the experimental approaches and the cost of measurement which makes many of them impractical solutions [19–21].

Among all of these techniques, the instrumented scratch test method is probably the most widespread technique for assessment of adhesion/cohesion parameters of a film–substrate system [22–25]. This traditional mechanical method is based on the interaction of the diamond testing probe with the coated surface. A probe with a linearly increasing load is pulled across the measured surface, while depth change is continuously measured. A moving indenter induces stresses in the tested film, which in reaction deforms elastically and/or plastically. Despite the experimental/principle simplicity, the scratching process is very complex [26–29]. When the deformation is sufficient to overcome the cohesive bonds in the film, cracks emerge. When the adhesive forces between film and substrate are surpassed, delamination or chipping of the film occurs. The objective of the scratch test is to find the so called critical loads (L_C), which are the forces inducing the typical failure modes. The great advantage of the scratch test stems from its possibility of tailoring the experimental parameters (normal load, indenter size), allowing the position of the von Mises stress maximum to be specified and, in turn, the sensitivity of the test for specific depth regions to be optimized. With increased load, the position of the von Mises stress is shifted to the substrate and this is more pronounced for larger indenter radii [30]. The proper dimensioning of the test can be performed based on Finite Element Modelling (FEM) [31–36] or analytical models [30]. Localizing the von Mises stress maximum—which in fact is supposed to be higher than yield stress—to the film–substrate interface is recommended when adhesion is the main concern. Due to its universality, the scratch test has spread from research to industrial use. The scratch test has been proved to be an efficient method for evaluation of adhesion and cohesion properties of various types of coated systems including hard protective coatings [37–40], optical thin films [41–43], medicinal coatings [44], and so on. Besides, it has been shown that also other types of materials can be beneficially explored such as plasma spray coatings [45], where for example the cohesion of individual splats can be studied [46]. In general, the tribological properties, including wear, of any material [47] can be studied as shown for example by Sola et al. for surface modified steel [48–50].

The evaluation of the critical loads is traditionally performed through the microscopic observation of the residual scratch groove and the indenter depth change record. While both techniques can be adequate in some cases on their own, they are often and beneficially used in combination to obtain more precise results. Nevertheless, even their combination may fail in the accurate determination of the critical loads in specific cases [30]. As a result, misleading conclusions about the film's durability and, in turn, improper inputs into the service life estimation and optimization process can be deduced. For example, consider the case of an opaque durable film on a hard brittle substrate (e.g., silicon), that is often the scenario in the semiconductor industry. In this case the stresses introduced during the scratch test can induce film and/or substrate cohesion failure or the adhesion failure between both components. Especially in the last two cases, the failure type classification and critical load identification can be misleading. When cohesive strength of the film itself is strong enough, the film can endure even after the first initiation of adhesion failure or substrate cohesion failure until it breaks away from the substrate at the latter part of the scratch track. As a result, the point of the film detachment is incorrectly identified as initial failure mode in the measured film–substrate system.

One of the possible solutions to overcome the abovementioned shortcomings of the standard evaluation techniques may be detection of acoustic emission (AE) signals recorded during the scratch test [51]. In general, acoustic emissions are elastic waves that are emitted during cracking or other irreversible changes in a material's internal structure. Although the AE method was already introduced in the scratch test some decades ago, its broader use has been hampered by insufficient hardware, impractical solutions, or weak software support for analysis. To the best of our knowledge, the use of AE during mechanical testing of thin films has been restricted mainly to thicker films (more than 5 μm) [52] or high normal loads (more than 1 N) [52–54]. Besides, these tests were often performed with spherical indenters of large radii (more than 100 μm) [52,54]. As a result, the von Mises stress maxima were shifted to the substrate, that is, further from the film–substrate interface. The analytical potential of the AE method is however extremely high. Combined use of AE detection with traditional techniques creates an unmatched approach for not only the detection of the initial cracks in the tested system, but also for deeper analysis and understanding of the failure modes' dynamics and causality during the scratch test.

Although the SiC thin films and their doped modifications are very promising materials in many structural applications and their mechanical properties have been studied thoroughly [55–57], there is a lack of systematic information about their deformation response to the scratch loading and adhesion strength. The tribological properties of various types of SiC_xN_y films and coatings have been studied and compared [57–61]. However, in most cases the deformation mechanism or the failure dynamic were neglected.

In this paper a combined scratch test evaluation based on the synergic use of the AE signal record along with the confocal microscope visual observation of the residual groove and the analysis of depth change record is presented. Hard a-SiC and a-SiCN thin films with the Si/C~1 on brittle Si substrates are explored and used as a demonstration system. The dynamics of the films' gradual damage is revealed and the individual failure modes are distinguished.

2. Acoustic Emission Introduction

Acoustic emission (AE) is the phenomenon describing the emission of elastic waves generated by release of internally stored energy during abrupt permanent changes in materials. Significant AE accompanies the initiation and propagation of cracks in rather brittle materials, however weaker AE also accompanies sliding of dislocation (slip planes) during deformation of crystalline material [62,63].

AE sensors have been routinely employed at macro-level for monitoring of structural strength of constructions and pipelines. They have been used for health monitoring and detecting of cracks in large pressure vessels, for control of water cooling pipeline systems in nuclear power plants, monitoring corrosion of reinforced concrete, or as failure control of important structural points [64–70]. Acoustic emission based methods also found their way to biological research with a drought stress experiment on leaves of common European trees [71]. Although the use of AE methods at macro scale has been successfully proved to be a reliable tool, their application at smaller scales is still a challenging task in terms of construction and data interpretation. Nevertheless, AE may provide a large amount of unique information about the structural integrity of a material/device and is still a very live topic [72,73].

The implementation of an acoustic emission method in the evaluation of the scratch test has been accompanied by various technical solutions throughout history. First, AE sensors based on accelerometers were introduced in the 1980s. Even with a cut-off frequency of around 50 kHz they were strongly disturbed by ambient mechanical vibrations, such as footsteps, motors' movement, and surface roughness during the tip move. This technology has been overcome by the utilization of the resonant AE sensors working at a frequency range of 30–1000 kHz due to the fact that the majority of the parasitic signals are rather low frequency signals below 50 kHz [74].

Localized stress fields and processes in a material during the scratch test results in acoustic emissions with frequencies up to 1 MHz, which leads to the demand for broadband piezoelectric detectors. The piezoelements are nowadays used in nearly all cases of modern AE systems. The piezoelectric response of used materials allows fabrication of compact sensors compatible with the majority of commercial scratch testers [75,76]. It should be noted that special sensors can be used even at elevated temperatures up to 1150 °C [77].

Since a scratch test on thin film is a sensitive micro-technique that can be affected by its surroundings, specific demands are made on the experimental arrangement. During the test, a micrometer radius tip interacts with the tested thin film causing a localized deformation zone with a size of hundreds of nanometers. The appropriate position of the AE sensor near the emission zone during the scratch test is of the highest importance. Small sensors can be placed on the mount of the scratching tip [75,78,79] or glued to the surface of sample [80,81]. Both solutions are often used in practice and both possess some advantages and disadvantages. A sensor attached to the tip benefits from its proximity to the material interaction volume from where the AE signal is emitted. However, the disadvantage is a small contact surface between the tip and the film's surface, which eventually transmits only a minor part of the signal that spreads omnidirectionally from the point of origin. Also the fixing procedure of the sensor itself is an issue, as its weight can interfere with the correct movement of the tip. In addition, such mounting makes manipulation during tip's change more difficult. Alternatively, the AE sensor can be directly placed on the sample surface. This approach benefits from its high sensitivity and easy implementation on pre-existing experimental setups. It should be noted that this is a preferable approach nowadays. The mounting position must be chosen appropriately as contact of the sensor with the tip must be avoided. On the other hand, the contamination of sample surface with the used adhesive is undesirable and may be considered as one of the biggest drawbacks. Since both abovementioned approaches possess substantial shortcomings, the use of AE methods in scratch testing has been only marginal and mostly devoted only to very specific case studies, some of which are listed below.

For example, research on AlN coated stainless steel by Choudhary [75] was focused on the use of the AE method as an indicator of a coating's cracking during the scratch test without use of any complex signal analysis. Similar use of the method concerning only the amplitude of the AE signal as a signalization of changes in material was used in bone toughness analysis by Kataruka [82] and research on several types of ceramics coatings (TiB_2, TiN, Al_2O_3-SiO_2-Cr_2O_3, CrN) by Ishikawa [83], Bhansali [84], Sekler [85], and Jensen [51].

A more sophisticated approach focusing on energy (wavelet transform) rather than amplitude of the AE signal was adopted in Piotrkovsky's research related to scratch test behavior of commercial galvanized coatings [86,87], grit scratch tests on aerospace alloys by Griffin [88–90], as well as a different type of tests on reinforced concrete by Sagasta, Zitto, and Piotrkovsky [91,92]. Research on deformation processes during scratch tests of molybdenum and tungsten by Zhou [93] also uses analysis of more complex outputs of the AE record. The deformation response of TiN thin films and the underlying Si substrate under sliding indentation was investigated by Benayoun [53] or Shiwa [94]. To reveal the substrate deformation, the film was chemically etched [94]. The combination of Focused Ion Beam (FIB) cross-sectioning of the residual groove and the AE was used to study the effect of substrate surface finish on the scratch induced failure mechanism of TiN on WC-Co substrate by Yang et al. [54].

It is not surprising, that the amount of data obtained from AE is massive. One of the solutions to handle the analysis of such an amount of data is the use of artificial neural networks [70,90].

Acoustic Emission Signal

The amplitude and frequency of AE events are functions of sample's material properties and geometry, and also depend on the source of AE. The complete record of the AE signal represents a large amount of data that may be inappropriately sizeable and too detailed for basic determination of critical events representing various types of film–substrate failure. It is more common to use a time compressed AE record, a so-called *Envelope* (see Figure 1a), that can be directly compared with the depth change record and the microscopic image of residual scratch groove. To visualize the dynamic of gradual damage of the tested material, one can choose the amplitude threshold and trace the number exceeding in full time resolution in the form of *Cumulative counts* (see Figure 1b). A full-resolution record of a specific AE event corresponding to the failure event occurring during the scratch test is called the *Hit* (see Figure 1c). It is defined by the amplitude and time thresholds. Using the Fourier transformation, a *Hit's spectrum* can be obtained (see Figure 1d). In addition to the hit's shape, its frequency spectra may be considered as a possible "fingerprint" of a particular type of failure mode in the film–substrate system [63].

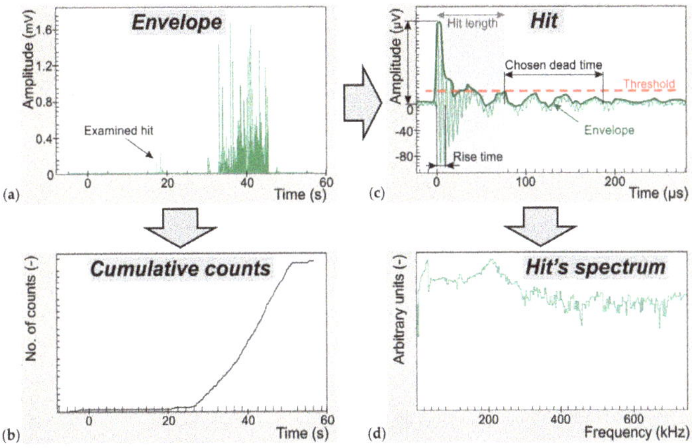

Figure 1. Parameters obtained from an acoustic emission (AE) analysis. (**a**) Compressed signal of complete AE record in form of the *Envelope* with (**b**) tracked exceedances of chosen amplitude threshold in form of the *Cumulative counts*. (**c**) Full time resolution AE event called *Hit* with its threshold described and (**d**) exported *Spectrum*.

3. Instrumentation and Methods

3.1. Acoustic Emission Setup

The acoustic emissions were detected using the ZEDO system (Dakel, Prague, Czech Republic) with the 10 MHz signal sampling rate allowing detections of AE signal variations with sub-microsecond resolution. The high dynamic range of the ZEDO system brings the possibility to register and analyze both the strong AE signals (complete breakage of the film) as well as weak AE signals originating from the small initial cracking. The AE record synchronized with the depth-change record allows accurate assignment of individual AE events to the microscopically observable failures in the residual scratch or the events in depth change record.

The samples were loaded on the special AE sample holder of our own design with the inbuilt high-sensitivity piezoelectric sensor and the dedicated preamplifier. Its design reflects the needs of high sensitivity and universal applicability and addresses the need of simple sample manipulation and avoiding the surface contamination by adhesive. Samples can be easily fixed to the holder using traditional cold or warm bonding (glue or wax, respectively).

3.2. Scratch Test Setup

The scratch tests were performed using the fully calibrated NanoTest instrument (MicroMaterials Ltd., Wales, UK) with the diamond cono-spherical Rockwell indenter with an actual radius of 9 μm. Samples were fixed on the special AE holder using a low temperature wax. In fact, the AE holder resembles the regular holder, except for the cable transferring the electric signal to the AE main unit (see Figure 2).

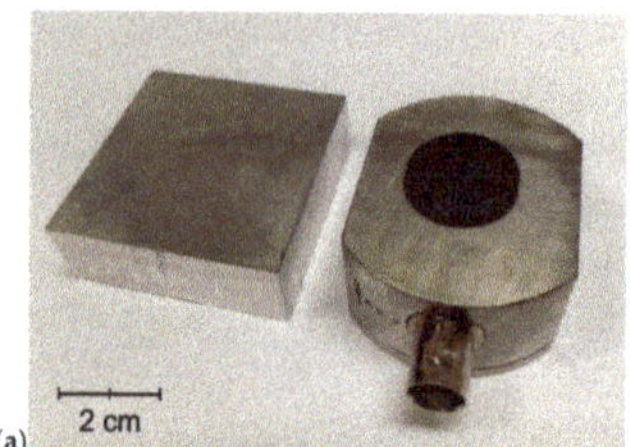

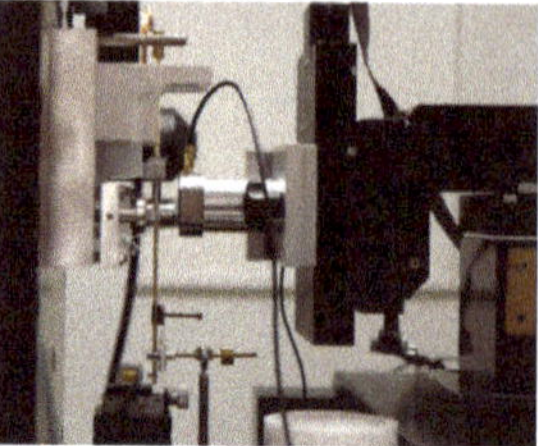

Figure 2. (**a**) Comparison of sample holders: traditional passive 5 cm × 5 cm specimen holder from duralumin (left) and the AE sample holder with built-in acoustic emission sensor and pre-amplifier (right). (**b**) Implementation of the AE holder into the NanoTest instrument.

During the ramped-load scratch tests the load on the indenter was linearly increased until the maximum of 500 mN. The experiments were performed as a 3-pass test, where the initial topography pass with very low topography load of 0.02 mN was followed by on-load scratch, after which came the final topography pass scanning the worn profile. The residual scratch groove was thoroughly explored using the laser scanning confocal microscope OLS LEXT 3100 (Olympus, Tokyo, Japan). The cross-sections of the residual scratch groove were made using the FEI Quanta 3D (Hillsboro, OR, USA) Dual-Beam SEM/FIB system with a Ga^+ ion source. Each test was repeated at least five times for sufficient data statistics and sample homogeneity. The evaluation of the instrumented scratch test was carried out based on a combination of three methods: (i) the depth change record, (ii) residual scratch imaging, and (iii) the acoustic emissions record.

4. Advanced Scratch Test Evaluation

The strength of the simultaneous AE detection during the scratch test supplementing the traditional methods will be demonstrated for a-SiC_xN_y ($y \geq 0$) films magnetron sputtered on silicon substrates. These SiC and SiCN films, depending on actual structure and composition, can possess high hardness [95], resistance to chemicals and abrasion even at high temperatures [96,97], as well high radiation resistance [98]. This excellent combination of physical and chemical properties makes them a very promising candidate for protective high temperature coatings [56,99,100]. It is especially the doping capability that allows tailoring of physical and material properties. For example, the imposed brittleness of the pure SiC can be partially suppressed by incorporation of N into the SiC structure. Besides, the band gap can be tuned by N content [101]. Since the a-SiC and a-SiCN films are often used in structural applications, their durability (including adhesion and cohesion) is of the highest importance and hence it has been studied extensively even in situ at elevated temperatures [102–104].

4.1. Combined Approach—Dynamic Revealed

A typical example, where the traditional approach lead to the incorrect determination of critical loads, can be instructively presented on two scratch tests performed on the SiC thin film (see Figure 3). A microscopic image of the first test's residual groove shows gross spallation beginning at the point marked by number "2", see Figure 3a. Small cracks inside the residual groove were detected with difficulty using higher magnification (not shown here) in the earlier stages starting from the point labeled by number "1". However, the acoustic emission record (see Figure 3b) shows an abrupt amplitude increase in the 25th second, indicating that the very first cracking in the film–substrate system (L_{C1}) occurred sooner than expected from the microscopic observation (marked by number "1" in Figure 3a). Besides, analysis of the depth change record reveals, that point "2"—considered as the onset of the gross spallation—is determined incorrectly, see Figure 3c. While the red curve representing the final topography indicates onset of the film failure already around the 340th micrometer of the track, the blue on-loaded scratch curve abruptly drops later, around the 380th micrometer. Taking all of these pieces of information into account, the dynamic of the film damage can be revealed and the onsets of the film–substrate failures can be evaluated correctly. In this particular case, film spallation occurs in the 380th micrometer (now marked as L_{C2}) both in the forward direction and even partially backward until the point "2".

Similarly to the previous example, relying solely only on the microscopic analysis (Figure 3d) inevitably leads to the misleading assumption that large area film delamination occurred at the point marked "1" and spread until the point "2". However, the depth change record analysis shows different dynamics of the film's failure, see Figure 3f. It is important to note that on-load depth as well as final depth records must be considered. The course of blue curve (on-load), representing the evolution of the tip penetration depth with an increasing normal load, increases linearly with a load until a sudden drop to a depth of around 8 μm occurs at 320th micrometer of the track, also labeled as the point "2". On the other hand, the red curve representing final topography profile shows surprising information: the depth curve drops abruptly to a depth of two microns around the 180th micrometer (point marked "1") and remains steady until point "2". Based on previous findings, it can be correctly determined that film delamination occurs at the moment of reaching point "2" (now designated L_{C2}), where the film delaminates significantly in the reverse direction until point "1". The last piece of information for full understanding of the failure dynamics is provided by the acoustic emission record. A single peak can be seen around 17th second of the AE record caused by the single crack in the film (see Figure 3e, position corresponds with the point "1" from the microscopic image Figure 3d). Its origin stems from a local growth defect on the film surface actually revealed in the initial topography (see green curve in Figure 3f). The backward spallation of the film, which detached at point "2" (=L_{C2}), was therefore stopped by this crack. The most interesting part in AE record can be seen around 22nd second, where signal amplitude rises again. Subsequent comparison with the microscopic record (see Figure 3d) points out cracking initiated at the film–substrate interface, which was still at the time covered by the coating (as explained above). The onset of this substrate cracking is assigned to the critical load L_{C1}.

Since the position of L_{C1} from this second scratch actually matches the L_{C1} in the first scratch, it can be concluded that substrate cracking below the still adhering film is indeed the very first failure event in all tested film–substrate systems. All this was revealed only through the combined use of the three evaluation techniques of ramped scratch test.

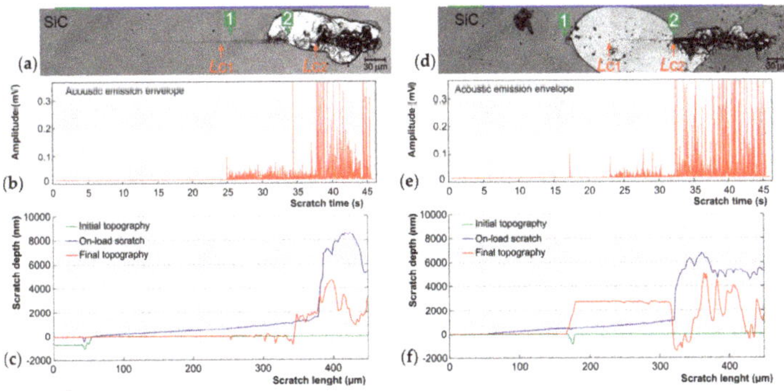

Figure 3. Advanced analysis of two ramped scratch tests on 00SiC sample with (**a**,**d**) the microscopic records, (**b**,**e**) the acoustic emissions records and (**c**,**f**) depth change records.

4.2. Extended Approach—Failure Type Revealed

The information content obtained from the scratch test and its reliability can be further enhanced via analysis of the residual groove by SEM-EDX mapping. The strength of such a complex approach can be manifested on the a-SiCN thin films annealed in air. It is well known that SiCN films with higher N content are prone to oxidation, especially at higher temperatures above 1000 °C, where the SiO_x

oxidation layer is formed [105]. This is another instructive example of how the sole visual observation may be misleading in terms of evaluation of scratch test.

Typical residual scratch groove for the SiCN film with the N content of 36 at % is shown in Figure 4. At the first glance, using only microscopic evaluation one may conclude that coating seems to be scratched through in the area highlighted by the violet rectangle, while the first critical load L_{C1} is practically undetectable even if combined with the depth change record. Subsequent detailed analysis based on the depth profiling of the 3D high resolution confocal microscope image and especially the electron microscope (see Figure 4d,e, respectively) revealed the penetration of the oxide SiO_x surface layer only (thus marked L_{Cox}), while the remaining unoxidized SiCN film remained intact. The exact point of scratching through the oxide layer was confirmed using electron microscopes' EDS mapping technique [106] as can be seen from Figure 4e, where the green arrow indicates the direction of the linear mapping profile inside the residual groove; the corresponding spectra are shown below. The blue dashed line of oxygen signal drops exactly at the point where SiO_x is scraped off. Another detail from the latter part of scratch track (see Figure 4d) shows the further evidence of the detached SiO_x layer even with its thickness evaluation. Perpendicular profile using calibrated confocal microscope shows a sharp transition between the SiO_x layer and the free surface of the SiCN film. Step measurement indicates thickness of SiO_x layer of t_0 = 100 nm. The coating itself proved to be unharmed since its plastic deformation at the deepest point of the residual groove is t_r = 400 nm, which still does not reach the full coating thickness of 2.6 μm. Moreover, the acoustic emission record clearly shows signal increase around 33th second. This is probably the very same failure mode caused by the film–substrate interface cracking, as described in the previous case on SiC samples (shown in Figure 3) and evidenced in the FIB-milled longitudinal cross-section. The lateral and median cracks are clearly identified in the film and substrate as can be seen in Figure 5. A detailed inspection of the SEM image shows no evidence of through-thickness cracking as all the cracks were stopped within the film. This may be related to the compressive stress in the films [56]. The initial cracks occurrence coincides well with the critical load L_{C1} obtain from the AE record. With the increasing load, the cracking in both the film as well as the substrate is more pronounced as shown on the perpendicular cross-section, see Figure 5.

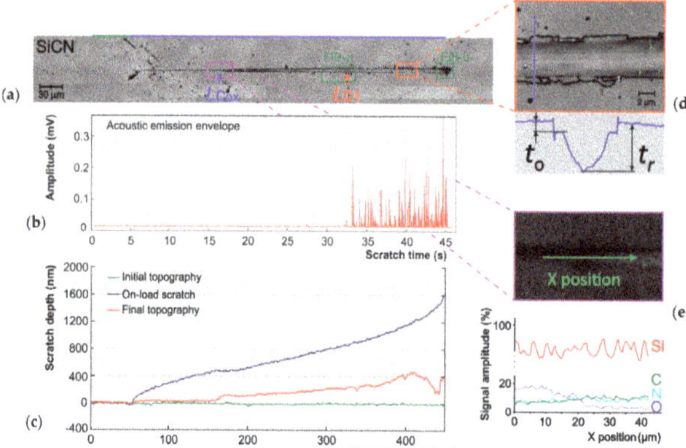

Figure 4. Detailed analysis of selected scratch test on sample SiCN with an oxidized surface using (**a**) microscopic record, (**b**) the acoustic emission record and (**c**) the depth change record. Detail (**d**) shows profile measurement of SiO_x oxide film thickness and detail (**e**) shows the oxide SiO_x film first penetration using the SEM-EDS linear mapping.

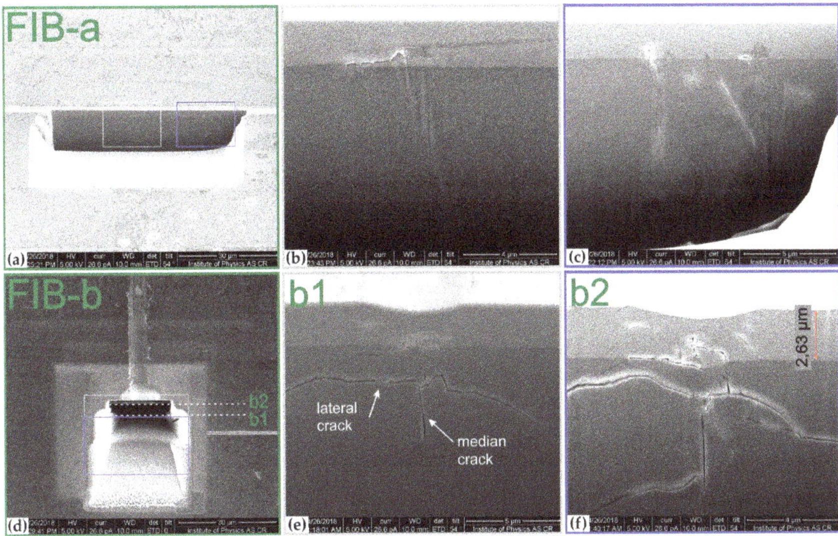

Figure 5. SEM images of the Focused Ion Beam (FIB) milled on the SiCN film for the(**a**) longitudinal cross section in the area of L_{c1} with (**b**,**c**) details and (**d**) the perpendicular cross section of the final part of the residual scratch groove with (**e**,**f**) two successive cross-sections.

4.3. AE as a Tool for First Failure Detection

The strength of the AE method in terms of the very first failure detection during the scratch test is clearly manifested by the comparison of values of critical loads evaluated by in situ AE and visual observation methods for SiC and SiCN thin films, see Figure 6. The as-deposited as well as air annealed films (700, 900 and 1100 °C) were explored. The primary goal was to test the effect of annealing temperature on tribological properties of these films with potential use for high temperature protective purposes. Introduction of nitrogen into the SiC structure led to the improvement of scratch resistance of originally deposited and also annealed coatings. However, the main finding in the context of combined scratch test evaluation described in this article is that the sensitivity of acoustic emissions exceeded the visual microscopic method in all cases. Critical loads obtained by the AE method were always lower (=detected sooner) than the microscopic ones. In extreme case the difference in the critical load values reaches 22%.

The use of the AE method as a complementary technique has been proved to be an important step forward in enhancement of the reliability and accuracy of the scratch test evaluation. However, the AE may go far beyond just identification of the very first cracking. The high sampling frequency of modern AE hardware units allows the recording of the AE signal with sub-microsecond resolution. As a result, the individual AE hits representing the particular physical events (cracking, chipping, delamination, etc.) in the film–substrate system can be subjected to a thorough analysis in time as well as frequency domain. Applying such an enhanced approach, it is possible to distinguish between various types of failures. Figure 7 illustrates the difference between the hit representing cracking in the SiC film only occurring at low load and the combined SiC film and underlying Si substrate cracking. The corresponding frequency spectra confirm the different response of the film–substrate system at different loads. Since various types of material failure are frequency dependent it is possible to online monitor and even classify the material/device/component deterioration that further emphasizes the beneficial role of AE method.

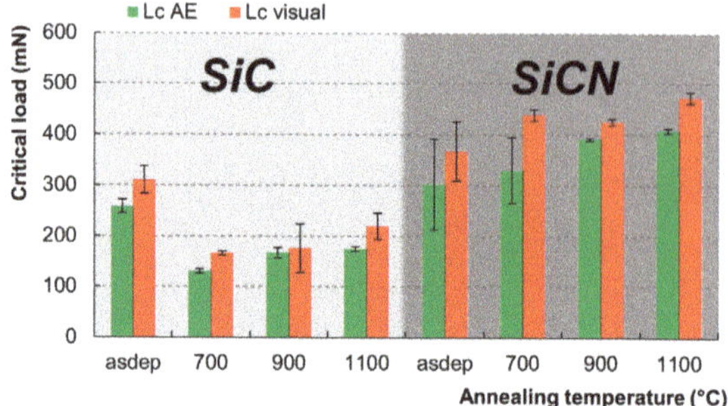

Figure 6. Sum of critical loads of the tested set of samples obtained through the acoustic emission evaluation and the traditional microscopic evaluation of residual scratch. Results show that AE method detected initial failure sooner in all cases.

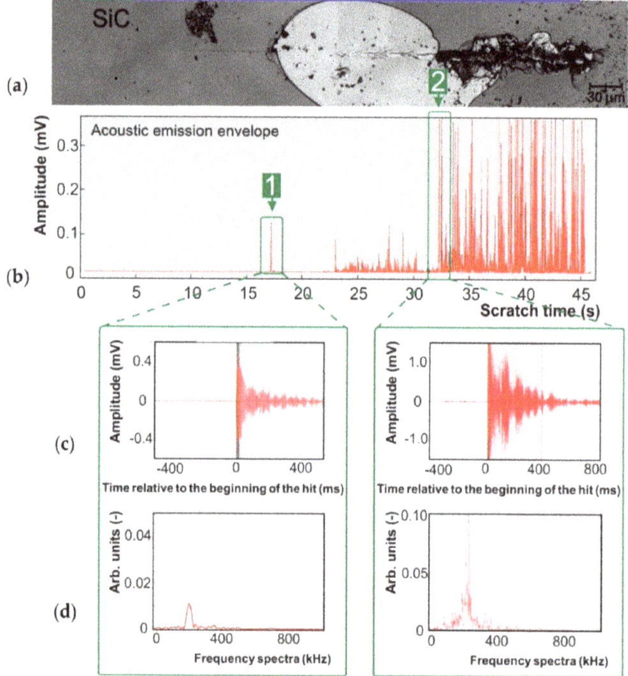

Figure 7. Detailed evaluation of specific observed AE hits. Firstly the microscopic record (**a**) was compared with the envelope of AE record (**b**). Then detailed evaluation in form of hit visualization (**c**) and hit's spectrum (**d**) of two AE events belonging to the lone film crack (marked "1") and jointly cracking of substrate and film (marked "2"). Note the different axis scale in both spectra (**d**).

5. Conclusions

The importance of the combined evaluation of the scratch tests on thin films was presented. In this combined approach, the standardly used visual observation of the residual groove and the depth change record were correlated with the simultaneous acoustic emission (AE) record. The illustrative examples were presented for the SiC and SiCN thin films deposited on Si substrates.

It was shown that proper analysis of the AE signal and its correlation with standard techniques can significantly increase the informative value of the scratch test. As a result, more reliable conclusions about the material response to external loading can be drawn. The present combined scratch/acoustic emission test evaluation has provided an excellent insight into the physical behaviour of SiC and SiCN films on silicon substrates during the single asperity scratching contact simulated by the diamond indenter.

The AE method provides unique information that enables the elucidation of the dynamics of the material failure with the potential to differentiate between the film and substrate cracking even at nano/micro scale. The use of AE in scratch testing is especially beneficial in the case of nontransparent thin films, where the AE signal can reveal subsurface cracking. Ignoring and/or neglecting the internal cracking may dramatically affect the scratch test results/findings. When they are used as input parameters into finite element modeling then the simulation findings can be distorted. Besides, it was shown that AE always detects critical loads sooner than visual observation.

Author Contributions: R.C. and J.T. conceived and designed the experiments, J.T. focused on scratch test experiments—their design, performance, and evaluation. R.C. and M.D. focused on crosschecking of acoustic emission data and advance Hit and Hit's spectra analysis. J.M. performed the FIB milling and SEM. J.T. and R.C. wrote the paper.

Acknowledgments: The authors gratefully acknowledge the support by the Operational Programme Research, Development and Education—European Regional Development Fund, project no CZ.02.1.01/0.0/0.0/16_019/0000754 of the Ministry of Education, Youth and Sports of the Czech Republic and the project TH03020245 of the Technology Agency of the Czech Republic, J.T. also acknowledges the Internal Grant of Palacky University (IGA_PrF_2018_009). J.M. acknowledges the support of the project LM2015087 of the Czech Ministry of Education, Youth and Sports.

Conflicts of Interest: The authors declare no conflict of interest.

References

1. Larsson, M.; Hollman, P.; Hedenqvist, P.; Hogmark, S.; Wahlström, U.; Hultman, L. Deposition and microstructure of PVD TiN—NbN multilayered coatings by combined reactive electron beam evaporation and DC sputtering. *Surf. Coat. Technol.* **1996**, *86–87 Pt 1*, 351–356. [CrossRef]
2. Reddy, I.N.; Reddy, V.R.; Sridhara, N.; Rao, V.S.; Bhattacharya, M.; Bandyopadhyay, P.; Basavaraja, S.; Mukhopadhyay, A.K.; Sharma, A.K.; Dey, A. Pulsed rf magnetron sputtered alumina thin films. *Ceram. Int.* **2014**, *40*, 9571–9582. [CrossRef]
3. Kamata, K.; Maeda, Y.; Yasui, K.; Moriyama, M. Preparation of Si_3N_4—SiC films by plasma CVD. *Int. J. High Technol. Ceram.* **1986**, *2*, 236. [CrossRef]
4. Kleps, L.; Caccavale, F.; Brusatin, G.; Angelescu, A.; Armelao, L. LPCVD silicon carbide and silicon carbonitride films using liquid single precursors. *Vacuum* **1995**, *46*, 979–981. [CrossRef]
5. Matthews, A.; Leyland, A. Developments in PVD tribological coatings. In Proceedings of the 5th ASM Heat Treatment and Surface Engineering Conference in Europe, Gothenburg, Sweden, 7–9 June 2000; p. 12.
6. Bescher, E.; Mackenzie, J. Sol-gel coatings for the protection of brass and bronze. *J. Sol-Gel Sci. Technol.* **2003**, *26*, 1223–1226. [CrossRef]
7. Kozuka, H.; Almeida, R.M.; Sakka, S. *Handbook of Sol-Gel Science and Technology Processing Characterization and Applications V. I: Sol-gel Processing*; Springer: New York, NY, USA, 2005.
8. Zajícová, V.; Exnar, P.; Staňová, I. Properties of hybrid coatings based on 3-trimethoxysilylpropyl methacrylate. *Ceramics-Silikáty* **2011**, *55*, 221–227.
9. Kuo, D.-H.; Yang, D.-G. Plasma-enhanced chemical vapor deposition of silicon carbonitride using hexamethyldisilazane and nitrogen. *Thin Solid Films* **2000**, *374*, 92–97. [CrossRef]
10. American Society for Testing and Materials. *ASTM D907-15 Standard Terminology of Adhesives*; American Society for Testing and Materials: West Conshohocken, PA, USA, 2015.
11. Pulker, H.K.; Perry, A.J.; Berger, R. Adhesion. *Surf. Technol.* **1981**, *14*, 25–39. [CrossRef]
12. Mittal, K.L. *Adhesion Measurement of Films and Coatings*; VSP: Utrecht, The Netherlands, 1995.
13. Chapman, B.N. Thin-film adhesion. *J. Vac. Sci. Technol.* **1974**, *11*, 106–113. [CrossRef]
14. Chopra, K.L. *Thin Film Phenomena*; McGraw-Hill: New York, NY, USA, 1969.

15. Campbell, D.S. *Handbook of Thin Film Technology*; Maissel, L.I., Glang, R., Eds.; McGraw-Hill: New York, NY, USA, 1970.
16. Jacobsson, R. Measurement of the adhesion of thin films. *Thin Solid Films* **1976**, *34*, 191–199. [CrossRef]
17. Katz, G. Adhesion of copper films to aluminum oxide using a spinel structure interface. *Thin Solid Films* **1976**, *33*, 99–105. [CrossRef]
18. Loh, R.L.; Rossington, C.; Evans, A.G. Laser technique for evaluating spall resistance of brittle coatings. *J. Am. Ceram. Soc.* **1986**, *69*, 139–142. [CrossRef]
19. Rickerby, D.S. A review of the methods for the measurement of coating-substrate adhesion. *Surf. Coat. Technol.* **1988**, *36*, 541–557. [CrossRef]
20. Valli, J. A review of adhesion test methods for thin hard coatings. *J. Vac. Sci. Technol. A Vac. Surf. Films* **1986**, *4*, 3007–3014. [CrossRef]
21. Chalker, P.R.; Bull, S.J.; Rickerby, D.S. A review of the methods for the evaluation of coating-substrate adhesion. *Mater. Sci. Eng. A* **1991**, *140*, 583–592. [CrossRef]
22. Sinha, S.K. 180 years of scratch testing. *Tribol. Int.* **2006**, *39*, 61. [CrossRef]
23. Bull, S.J. Techniques for improving thin film adhesion. *Vacuum* **1992**, *43*, 517–520. [CrossRef]
24. Laugier, M. The development of the scratch test technique for the determination of the adhesion of coatings. *Thin Solid Films* **1981**, *76*, 289–294. [CrossRef]
25. Burnett, P.J.; Rickerby, D.S. The relationship between hardness and scratch adhession. *Thin Solid Films* **1987**, *154*, 403–416. [CrossRef]
26. Laugier, M.T. An energy approach to the adhesion of coatings using the scratch test. *Thin Solid Films* **1984**, *117*, 243–249. [CrossRef]
27. Burnett, P.J.; Rickerby, D.S. The scratch adhesion test: An elastic-plastic indentation analysis. *Thin Solid Films* **1988**, *157*, 233–254. [CrossRef]
28. Bull, S.J.; Rickerby, D.S.; Matthews, A.; Leyland, A.; Pace, A.R.; Valli, J. The use of scratch adhesion testing for the determination of interfacial adhesion: The importance of frictional drag. *Surf. Coat. Technol.* **1988**, *36*, 503–517. [CrossRef]
29. Kendall, K. The adhesion and surface energy of elastic solids. *J. Phys. D Appl. Phys.* **1971**, *4*, 1186. [CrossRef]
30. Schwarzer, N.; Duong, Q.H.; Bierwisch, N.; Favaro, G.; Fuchs, M.; Kempe, P.; Widrig, B.; Ramm, J. Optimization of the scratch test for specific coating designs. *Surf. Coat. Technol.* **2011**, *206*, 1327–1335. [CrossRef]
31. Hegadekatte, V.; Huber, N.; Kraft, O. Finite element based simulation of dry sliding wear. *Model. Simul. Mater. Sci. Eng.* **2005**, *13*, 57. [CrossRef]
32. Hegadekatte, V.; Huber, N.; Kraft, O. Modeling and simulation of wear in a pin on disc tribometer. *Tribol. Lett.* **2006**, *24*, 51. [CrossRef]
33. Steiner, L.; Bouvier, V.; May, U.; Hegadekatte, V.; Huber, N. Modelling of unlubricated oscillating sliding wear of DLC-coatings considering surface topography, oxidation and graphitisation. *Wear* **2010**, *268*, 1184–1194. [CrossRef]
34. Holmberg, K.; Ronkainen, H.; Laukkanen, A.; Wallin, K.; Erdemir, A.; Eryilmaz, O. Tribological analysis of TiN and DLC coated contacts by 3D FEM modelling and stress simulation. *Wear* **2008**, *264*, 877–884. [CrossRef]
35. Holmberg, K.; Ronkainen, H.; Laukkanen, A.; Wallin, K. Friction and wear of coated surfaces—scales, modelling and simulation of tribomechanisms. *Surf. Coat. Technol.* **2007**, *202*, 1034–1049. [CrossRef]
36. Holmberg, K.; Laukkanen, A.; Ronkainen, H.; Wallin, K.; Varjus, S.; Koskinen, J. Tribological contact analysis of a rigid ball sliding on a hard coated surface: Part I: Modelling stresses and strains. *Surf. Coat. Technol.* **2006**, *200*, 3793–3809. [CrossRef]
37. Beake, B.D.; Vishnyakov, V.M.; Valizadeh, R.; Colligon, J.S. Influence of mechanical properties on the nanoscratch behaviour of hard nanocomposite SiN/Si_3N_4 coatings on si. *J. Phys. D Appl. Phys.* **2006**, *39*, 1392. [CrossRef]
38. Beake, B.D.; Endrino, J.L.; Kimpton, C.; Fox-Rabinovich, G.S.; Veldhuis, S.C. Elevated temperature repetitive micro-scratch testing of AlCrN, TiAlN and AlTiN PVD coatings. *Int. J. Refract. Met. Hard Mater.* **2017**, *69*, 215–226. [CrossRef]

39. Fox-Rabinovich, G.S.; Kovalev, A.I.; Aguirre, M.H.; Beake, B.D.; Yamamoto, K.; Veldhuis, S.C.; Endrino, J.L.; Wainstein, D.L.; Rashkovskiy, A.Y. Design and performance of AlTiN and TiAlCrN PVD coatings for machining of hard to cut materials. *Surf. Coat. Technol.* **2009**, *204*, 489–496. [CrossRef]
40. Bull, S.J.; Rickerby, D.S.; Robertson, T.; Hendry, A. The abrasive wear resistance of sputter ion plated titanium nitride coatings. *Surf. Coat. Technol.* **1988**, *36*, 743–754. [CrossRef]
41. Perry, A.J.; Pulker, H.K. Hardness, adhesion and abrasion resistance of TiO_2 films on glass. *Thin Solid Films* **1985**, *124*, 323–333. [CrossRef]
42. Guruvenket, S.; Azzi, M.; Li, D.; Szpunar, J.A.; Martinu, L.; Klemberg-Sapieha, J.E. Structural, mechanical, tribological, and corrosion properties of a-SiC:H coatings prepared by PECVD. *Surf. Coat. Technol.* **2010**, *204*, 3358–3365. [CrossRef]
43. Borrero-López, O.; Hoffman, M.; Bendavid, A.; Martin, P.J. Mechanical properties and scratch resistance of filtered-arc-deposited titanium oxide thin films on glass. *Thin Solid Films* **2011**, *519*, 7925–7931. [CrossRef]
44. Beake, B.D.; Liskiewicz, T.W. Comparison of nano-fretting and nano-scratch tests on biomedical materials. *Tribol. Int.* **2013**, *63*, 123–131. [CrossRef]
45. Das, D.K.; Srivastava, M.P.; Joshi, S.V.; Sivakumar, R. Scratch adhesion testing of plasma-sprayed yttria-stabilized zirconia coatings. *Surf. Coat. Technol.* **1991**, *46*, 331–345. [CrossRef]
46. Chen, S.-Y.; Ma, G.-Z.; Wang, H.-D.; He, P.-F.; Wang, H.-M.; Liu, M. Evaluation of adhesion strength between amorphous splat and substrate by micro scratch method. *Surf. Coat. Technol.* **2018**, *344*, 43–51. [CrossRef]
47. Brostow, W.; Lobland, H.E.H. *Materials: Introduction and Applications*; John Wiley & Sons: New York, NY, USA, 2017.
48. Sola, R. Post-treatment surface morphology effect on the wear and corrosion resistance of nitrided and nitrocarburized 41 CrAlMo7 steel. *La Metall. Ital.* **2010**, *5*, 21–31.
49. Sola, R.; Giovanardi, R.; Parigi, G.; Veronesi, P.; Berto, F. A novel methods for fracture toughness evaluation of tool steels with post-tempering cryogenic treatment. *Metals* **2017**, *7*, 75. [CrossRef]
50. Sola, R.; Poli, G.; Veronesi, P.; Giovanardi, R. Effects of surface morphology on the wear and corrosion resistance of post-treated nitrided and nitrocarburized 42CrMo4. *Metall. Mater. Trans. A* **2014**, *45*, 2827–2833. [CrossRef]
51. Jensen, H.; Jensen, U.M.; Sorensen, G. Reactively sputtered Cr nitride coatings studied using the acoustic emission scratch test technique. *Surf. Coat. Technol.* **1995**, *74–75*, 297–305. [CrossRef]
52. Yamamoto, S.; Ichimura, H. Effects of intrinsic properties of TiN coatings on acoustic emission behavior at scratch test. *J. Mater. Res.* **2011**, *7*, 2240–2247. [CrossRef]
53. Benayoun, S.; Hantzpergue, J.J.; Bouteville, A. Micro-scratch test study of TiN films grown on silicon by chemical vapor deposition. *Thin Solid Films* **2001**, *389*, 187–193. [CrossRef]
54. Yang, J.; Roa, J.; Odén, M.; Johansson-Jõesaar, M.P.; Esteve, J.; Llanes, L. Substrate surface finish effects on scratch resistance and failure mechanisms of TiN-coated hardmetals. *Surf. Coat. Technol.* **2015**, *265*, 174–184. [CrossRef]
55. Kulikovsky, V.; Vorlicek, V.; Ctvrtlik, R.; Bohac, P.; Suchanek, J.; Blahova, O.; Jastrabik, L. Mechanical and tribological properties of coatings sputtered from SiC target in the presence of CH_4 gas. *Surf. Coat. Technol.* **2011**, *205*, 3372–3377. [CrossRef]
56. Kulikovsky, V.; Ctvrtlik, R.; Vorlicek, V.; Zelezny, V.; Bohac, P.; Jastrabik, L. Effect of air annealing on mechanical properties and structure of SiC_xN_y magnetron sputtered films. *Surf. Coat. Technol.* **2014**, *240*, 76–85. [CrossRef]
57. Hoche, H.; Allebrandt, D.; Bruns, M.; Riedel, R.; Fasel, C. Relationship of chemical and structural properties with the tribological behavior of sputtered SiCN films. *Surf. Coat. Technol.* **2008**, *202*, 5567–5571. [CrossRef]
58. Zhou, Y.; Probst, D.; Thissen, A.; Kroke, E.; Riedel, R.; Hauser, R.; Hoche, H.; Broszeit, E.; Kroll, P.; Stafast, H. Hard silicon carbonitride films obtained by RF-plasma-enhanced chemical vapour deposition using the single-source precursor bis(trimethylsilyl)carbodiimide. *J. Eur. Ceram. Soc.* **2006**, *26*, 1325–1335. [CrossRef]
59. Probst, D.; Hoche, H.; Zhou, Y.; Hauser, R.; Stelzner, T.; Scheerer, H.; Broszeit, E.; Berger, C.; Riedel, R.; Stafast, H.; et al. Development of PE-CVD Si/C/N:H films for tribological and corrosive complex-load conditions. *Surf. Coat. Technol.* **2005**, *200*, 355–359. [CrossRef]
60. Wu, Y.; Zhu, S.; Zhang, Y.; Liu, T.; Rao, Y.; Luo, L.; Wang, Q. The adhesion strength and deuterium permeation property of sic films synthesized by magnetron sputtering. *Int. J. Hydrogen Energy* **2016**, *41*, 10827–10832. [CrossRef]

61. Bhattacharyya, A.S.; Mishra, S.K. Micro/nanomechanical behavior of magnetron sputtered Si–C–N coatings through nanoindentation and scratch tests. *J. Micromech. Microeng.* **2011**, *21*, 015011. [CrossRef]
62. Nazarchuk, Z.; Skalskyi, V.; Serhiyenko, O. *Acoustic Emission—Methodology and Application*; Springer International Publishing: Cham, Switzerland, 2017.
63. Scruby, C.B. An introduction to acoustic emission. *J. Phys. E Sci. Instrum.* **1987**, *20*, 946. [CrossRef]
64. Stulen, F.B.; Kiefner, J.F. Evaluation of acoustic emission monitoring of buried pipelines. In Proceedings of the 1982 Ultrasonics Symposium, San Diego, CA, USA, 27–29 October 1982; pp. 898–903.
65. Lim, J. In Underground pipeline leak detection using acoustic emission and crest factor technique. In *Advances in Acoustic Emission Technology*; Shen, G., Wu, Z., Zhang, J., Eds.; Springer: New York, NY, USA, 2015; pp. 445–450.
66. Bakirov, M.B.; Povarov, V.P.; Gromov, A.F.; Levchuk, V.I. Development of a technology for continuous acoustic emission monitoring of in-service damageability of metal in safety-related NPP equipment. *Nucl. Energy Technol.* **2015**, *1*, 32–36. [CrossRef]
67. Lee, J.-H.; Lee, M.-R.; Kim, J.-T.; Luk, V.; Jung, Y.-H. A study of the characteristics of the acoustic emission signals for condition monitoring of check valves in nuclear power plants. *Nucl. Eng. Des.* **2006**, *236*, 1411–1421. [CrossRef]
68. Minemura, O.; Sakata, N.; Yuyama, S.; Okamoto, T.; Maruyama, K. Acoustic emission evaluation of an arch dam during construction cooling and grouting. *Constr. Build. Mater.* **1998**, *12*, 385–392. [CrossRef]
69. Sagaidak, A.; Bardakov, V.; Elizarov, S.; Terentyev, D. The use of acoustic emission method in the modern construction. In Proceedings of the 31st Conference of the European Working Group on Acoustic Emission, Dresden, Germany, 31 September 2014.
70. Bonaccorsi, L.; Calabrese, L.; Campanella, G.; Proverbio, E. Artificial neural network analyses of AE data during long-term corrosion monitoring of a post-tensioned concrete beam. *J. Acoust. Emiss.* **2012**, *30*, 40–53.
71. Rosner, S. Acoustic emission related to drought stress response of four deciduous broad-leaved woody species. *J. Acoust. Emiss.* **2012**, *30*, 11–20.
72. Holford, K.M. Acoustic emission–basic principles and future directions. *Strain* **2008**, *36*, 51–54. [CrossRef]
73. Guo, Y.B.; Ammula, S.C. Real-time acoustic emission monitoring for surface damage in hard machining. *Int. J. Mach. Tools Manuf.* **2005**, *45*, 1622–1627. [CrossRef]
74. Julia-Schmutz, C.; Hintermann, H.E. Microscratch testing to characterize the adhesion of thin layers. *Surf. Coat. Technol.* **1991**, *48*, 1–6. [CrossRef]
75. Choudhary, R.K.; Mishra, P. Use of acoustic emission during scratch testing for understanding adhesion behavior of aluminum nitride coatings. *J. Mater. Eng. Perform.* **2016**, *25*, 2454–2461. [CrossRef]
76. Proctor, T.M. An improved piezoelectric acoustic emission transducer. *J. Acoust. Soc. Am.* **1982**, *71*, 1163–1168. [CrossRef]
77. Turner, R.C.; Fuierer, P.A.; Newnham, R.E.; Shrout, T.R. Materials for high temperature acoustic and vibration sensors: A review. *Appl. Acoust.* **1994**, *41*, 299–324. [CrossRef]
78. Von Stebut, J.; Lapostolle, F.; Bucsa, M.; Vallen, H. Acoustic emission monitoring of single cracking events and associated damage mechanism analysis in indentation and scratch testing. *Surf. Coat. Technol.* **1999**, *116–119*, 160–171. [CrossRef]
79. Hiroyuki, K.; Akio, M.; Ken, Y.; Shigeru, N.; Yasushi, F. Analysis of AE signals during scratch test on the coated paperboard. *J. Acoust. Emiss.* **2012**, *30*, 1–11.
80. Gallego, A.; Gil, J.F.; Vico, J.M.; Ruzzante, J.E.; Piotrkowski, R. Coating adherence in galvanized steel assessed by acoustic emission wavelet analysis. *Scr. Mater.* **2005**, *52*, 1069–1074. [CrossRef]
81. Gallego, A.; Gil, J.F.; Castro, E.; Piotrkowski, R. Identification of coating damage processes in corroded galvanized steel by acoustic emission wavelet analysis. *Surf. Coat. Technol.* **2007**, *201*, 4743–4756. [CrossRef]
82. Kataruka, A.; Mendu, K.; Okeoghene, O.; Puthuvelil, J.; Akono, A.-T. Microscopic assessment of bone toughness using scratch tests. *Bone Rep.* **2017**, *6*, 17–25. [CrossRef] [PubMed]
83. Ishikawa, H.; Ohnaka, T.; Hirohashi, M. Evaluation of acoustic emission generated in scratch testing of ceramic coatings. *J. Surf. Finish. Soc. Jpn.* **1994**, *45*, 296–300. [CrossRef]
84. Bhansali, K.J.; Kattamis, T.Z. Quality evaluation of coatings by automatic scratch testing. *Wear* **1990**, *141*, 59–71. [CrossRef]
85. Sekler, J.; Steinmann, P.A.; Hintermann, H.E. The scratch test: Different critical load determination techniques. *Surf. Coat. Technol.* **1988**, *36*, 519–529. [CrossRef]

86. Piotrkowski, R.; Castro, E.; Gallego, A. Wavelet power, entropy and bispectrum applied to AE signals for damage identification and evaluation of corroded galvanized steel. *Mech. Syst. Signal Process.* **2009**, *23*, 432–445. [CrossRef]
87. Piotrkowski, R.; Gallego, A.; Castro, E.; García-Hernandez, M.T.; Ruzzante, J.E. Ti and Cr nitride coating/steel adherence assessed by acoustic emission wavelet analysis. *NDT E Int.* **2005**, *38*, 260–267. [CrossRef]
88. Griffin, J.M.; Torres, F. Dynamic precision control in single-grit scratch tests using acoustic emission signals. *Int. J. Adv. Manuf. Technol.* **2015**, *81*, 935–953. [CrossRef]
89. Griffin, J. Traceability of acoustic emission measurements for a proposed calibration method—Classification of characteristics and identification using signal analysis. *Mech. Syst. Signal Process.* **2015**, *50–51*, 757–783. [CrossRef]
90. Griffin, J.; Chen, X. Classification of the acoustic emission signals of rubbing, ploughing and cutting during single grit scratch tests. *Int. J. Nanomanuf.* **2006**, *1*, 189–209. [CrossRef]
91. Sagasta, F.; Zitto, M.E.; Piotrkowski, R.; Benavent-Climent, A.; Suarez, E.; Gallego, A. Acoustic emission energy *b*-value for local damage evaluation in reinforced concrete structures subjected to seismic loadings. *Mech. Syst. Signal Process.* **2018**, *102*, 262–277. [CrossRef]
92. Zitto, M.; Scaramal, M.; Sagasta, F.; Piotrkowski, R.; Gallego, A.; Castro, E. AE signal processing in dynamical tests of reinforced concrete structures. In Proceedings of the 30th European Conference on Acoustic Emission Testing & 7th International Conference on Acoustic Emission, Granada, Spain, 12–15 September 2012.
93. Zhou, W.; He, Y.; Lu, X. Acoustic emission in scratch processes of metals. *Insight Non-Destr. Test. Cond. Monit.* **2015**, *57*, 635–642. [CrossRef]
94. Shiwa, M.; Weppelmann, E.R.; Bendeli, A.; Swain, M.V.; Munz, D.; Kishi, T. Acoustic emission and precision force-displacement observations of spherical indentations into TiN films on silicon. *Surf. Coat. Technol.* **1994**, *68–69*, 598–602. [CrossRef]
95. Badzian, A.; Badzian, T.; Drawl, W.D.; Roy, R. Silicon carbonitride: A rival to cubic boron nitride. *Diam. Relat. Mater.* **1998**, *7*, 1519–1525. [CrossRef]
96. An, L.; Riedel, R.; Konetschny, C.; Kleebe, H.J.; Raf, R. Newtonian viscosity of amorphous silicon carbonitride at high temperature. *J. Am. Ceram. Soc.* **1998**, *81*, 1349–1352. [CrossRef]
97. Riedel, R.; Kleebe, H.J.; Schonfelder, H.; Aldinger, F. A covalent micro nanocomposite resistant to high-temperature oxidation. *Nature* **1995**, *374*, 526–528. [CrossRef]
98. Lebedev, A.A.; Ivanov, A.M.; Strokan, N.B. Radiation resistance of SiC and nuclear-radiation detectors based on SiC films. *Semiconductors* **2004**, *38*, 125–147. [CrossRef]
99. Kulikovsky, V.; Vorlíček, V.; Boháč, P.; Stranyánek, M.; Čtvrtlík, R.; Kurdyumov, A. Mechanical properties of amorphous and microcrystalline silicon films. *Thin Solid Films* **2008**, *516*, 5368–5375. [CrossRef]
100. Kulikovsky, V.; Vorlíček, V.; Boháč, P.; Stranyánek, M.; Čtvrtlík, R.; Kurdyumov, A.; Jastrabik, L. Hardness and elastic modulus of amorphous and nanocrystalline SiC and Si films. *Surf. Coat. Technol.* **2008**, *202*, 1738–1745. [CrossRef]
101. Yang, J. A harsh environment wireless pressure sensing solution utilizing high temperature electronics. *Sensors* **2013**, *13*, 2719–2734. [CrossRef] [PubMed]
102. Riedel, R.; Kienzle, A.; Dressler, W.; Ruwisch, L.; Bill, J.; Aldinger, F. A silicoboron carbonitride ceramic stable to 2000 °C. *Nature* **1996**, *382*, 796–798. [CrossRef]
103. Raj, R.; An, L.; Shah, S.; Riedel, R.; Fasel, C.; Kleebe, H.-J. Oxidation kinetics of an amorphous silicon carbonitride ceramic. *J. Am. Ceram. Soc.* **2001**, *84*, 1803–1810. [CrossRef]
104. Ctvrtlik, R.; Al-Haik, M.; Kulikovsky, V. Mechanical properties of amorphous silicon carbonitride thin films at elevated temperatures. *J. Mater. Sci.* **2015**, *50*, 1553–1564. [CrossRef]
105. Ctvrtlik, R.; Kulikovsky, V.; Vorlicek, V.; Tomastik, J.; Drahokoupil, J.; Jastrabik, L. Mechanical properties and microstructural characterization of amorphous SiC_xN_y thin films after annealing beyond 1100 °C. *J. Am. Ceram. Soc.* **2016**, *99*, 996–1005. [CrossRef]
106. Goldstein, J.; Newbury, D.E.; Joy, D.C.; Lyman, C.E.; Echlin, P.; Lifshin, E.; Sawyer, L.; Michael, J.R. *Scanning Electron Microscopy and X-ray Microanalysis*, 3rd ed.; Springer Science + Business Media: New York, NY, USA, 2003; p. 689.

Article

Raman Spectroscopy for Reliability Assessment of Multilayered AlCrN Coating in Tribo-Corrosive Conditions

Janis Baronins [1], Maksim Antonov [1,*], Sergei Bereznev [2], Taavi Raadik [2,3] and Irina Hussainova [1]

1 Department of Mechanical and Industrial Engineering, Tallinn University of Technology, 12616 Tallinn, Estonia; Janis.Baronins@ttu.ee (J.B.); Irina.Hussainova@ttu.ee (I.H.)

2 Department of Materials and Environmental Technology, Tallinn University of Technology, 12616 Tallinn, Estonia; Sergei.Bereznev@ttu.ee (S.B.); Taavi.Raadik@ttu.ee (T.R.)

3 European Space Agency ESA-ESTEC, Materials and Processes Section, 2201 Noordwijk, The Netherlands

* Correspondence: Maksim.Antonov@ttu.ee; Tel.: +372-620-335-5

Received: 18 May 2018; Accepted: 20 June 2018; Published: 26 June 2018

Abstract: In this study, a multilayered AlCrN coating has been employed as a protective layer for steel used in tribo-corrosive conditions. The coating was deposited by a lateral rotating cathode arc PVD technology on a AISI 316L stainless steel substrate. A ratio of Al/(Al + Cr) was varied from 0.5 up to 0.6 in the AlCrN layer located above Cr adhesion and gradient CrN interlayers. A Raman spectroscopy and potentiodynamic polarization scan were used to determine the resistance in tribo-corrosive (3.5 wt % NaCl) conditions. Correlation between sliding contact surface chemistry and measured tribological properties of material was supported with static corrosion experiments. The corrosion mechanisms responsible for surface degradation are reported.

Keywords: multilayered AlCrN coating; Raman spectroscopy; tribo-corrosion; sliding wear

1. Introduction

One of the problems in the industrial application of moving bodies concerns the mechanical interaction between sliding surfaces and surface chemical reactions or corrosion occurring in reactive environments such as an aqueous media. A simplified description of tribo-corrosion phenomenon is related to a material transformation process due to simultaneous corrosion and wear taking place at contacting surfaces in relative motion [1].

Corrosion resistance is one of the most important factors to be taken into consideration for manufacturing metal products, as the formation of rust can have a devastating impact on the performance. Protection of the metal surfaces with physical vapour deposited coatings is a widely used technique. It could be assumed that such treatment will be even more relevant in future due to boosting of 3D metal printing technologies (additive manufacturing) [2]. Reliable lifetime prediction for a component used in an aqueous corrosive environment requires the identification of corrosion failure modes. Such failure modes can be pitting (if halide ions are present), stress-corrosion caused cracking by hydrogen embrittlement and corrosion fatigue [3].

The effect of mechanical stimulation on chemical degradation of materials and, vice-versa, the influence of corrosion on the mechanical response of contacting materials are of great concern for modern technologies including power generation, marine, and offshore industries. Materials properties, surface transformations, and electrochemical reactions are important aspects to be considered during materials selection for any specific application as cumulative effects of mechanical and chemical factors can result in unexpected behaviour and catastrophic loss of integrity. However, the chemo-mechanical mechanisms of tribo-corrosion are not yet well-understood and are extremely complex as they involve a great number of parameters [1,4–6]. A realistic evaluation of materials reliability is further hindered by the experimental difficulties in process characterization. Moreover, the overall rate of material degradation is rarely the sum of just corrosion and wear but is influenced by multiple reactions and transformations that take place during tribo-corrosive interactions. Therefore, an attempt to use Raman spectroscopy as a non-destructive and relatively fast method for understanding processes of tribo-corrosion is of potential benefit [7].

Nowadays the use of protective coatings containing carbon, oxygen, or nitrogen (e.g., carbides, nitrides, carbonitrides, or oxynitrides) is considered to be a practical method for improvement of the performance of metals and alloys [8–10]. Transition metal nitrides ensure the high hardness, acceptable wear, and corrosion resistance when applied as physical vapour deposited (PVD) coatings to enable application under aggressive environments [11,12]. Dominating phase transition changes from cubic to hexagonal have been found in $Al_xCr_{1-x}N$ by increasing x up to about 0.71 [13]; however, this value has not been strictly defined.

In many cases, a ceramic coating cannot be applied directly to an SS substrate due to insufficient bonding efficiency. As the result of this, intensive delamination of a coating can take place. It is especially harmful, if emission of Cr containing particles takes place that can oxidize into a toxic and cancerogenic Cr(VI) [14]. Accordingly, an adhesive interlayer with as possible similar lattice parameters can be applied. In such situation, a process of inter-diffusion between coating and substrate may occur. Cohesive energy densities or solubility parameters should match according to thermodynamic considerations to attain good bonding between a substrate and an adhesive layer [15]. A combination of Fe and Cr satisfy these conditions as both have bcc structures. It is well known that specific interactions between the components enable blending the miscible materials [16]. The ideal work of adhesion properties of the Cr(100)/Fe(100) and Cr(110)/Fe(110) abrupt interfaces has been predicted to be about 5.4 $J \cdot m^{-2}$. Endothermic intermixing occurs at the interface of Cr film and Fe substrate, exhibiting a very strong adhesion caused by strong covalent bonding in addition to metallic cohesion and nearly lack of strain [14]. Intermixing causes a favourable concentration gradient transition zones distinguished by thermodynamic compatibility of a substrate-coating system [17].

Herein, the tribo-corrosive processes occurring at multilayered AlCrN PVD coatings deposited over stainless steel (SS) substrate demonstrating an applicability of Raman spectroscopy for determination of corrosion products and possible coating failures under static and tribologically initiated conditions is reported.

2. Materials and Methods

2.1. Materials

Austenitic conventional (produced by casting and rolling) SS AISI 316L (UNS S31603, dimension: 25 mm × 15 mm × 5 mm) supplied by *Outokumpu* (Helsinki, Finland) was used as the substrate material for the compositions described. Selected Fe based SS typically contains Cr (17.2 wt %) C (0.02 wt %), Ni (10.1 wt %) and Mo (2.1 wt %) according to properties provided by the producer. A small concentration of several other elements like Si, P, S, Mn, and N can be detected during elemental analysis.

An arithmetical mean roughness of the substrates $R_a \leq 0.02$ µm was reached using a Phoenix 4000 (*Buehler*, Lake Bluff, IL, USA) polishing system by applying SiC papers (Buehler) down to grade P4000 (*MicroCut S*, *Buehler*). Substrates were cleaned with an isopropanol for 50 min in an ultrasonic bath and then sputter-cleaned in a chamber with argon plasma with the bias voltage of 850 V at 425 °C for 1 h. Chromium adhesive and gradient CrN interlayers were used in order to provide sufficient adhesion of the AlCrN coating to the substrate. The structure of alternating layers with an Al/(Al + Cr) ratio of 0.6 and 0.5 was produced by varying the Cr cathode arc current, [5]. The adhesion of the coating to the substrate was characterised as class 1 according to VDI 3198. The thickness of the AlCrN coating was established as 3 µm and the thickness of the Cr/CrN interlayer was about 0.3 µm (Kalotest method by

BAQ GmbH KaloMAX ball cratering device according to EN1071-2007). The deposition temperature for the substrate was 450 °C. The schematic representation of the coating described is given in Figure 1a. A dominating cubic CrN phase was confirmed by XRD in the structure of the AlCrN PVD coating after deposition, as demonstrated in Figure 1b (minerals with similar XRD signals are indicated). The main properties of the coating are listed in Table 1.

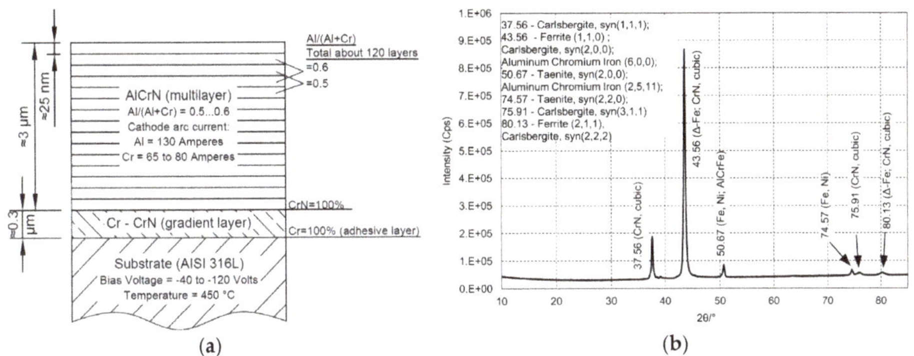

Figure 1. A schematic illustration of the multilayered AlCrN PVD coating on the stainless steel substrate (**a**); and XRD (made by *Rigaku Ultima IV*, Tokyo, Japan) diffractogram of AlCrN PVD coating deposited onto stainless steel AISI 316L substrate, indicating a dominating cubic CrN phase in the coating (**b**).

Counter-body balls of yttria-stabilized tetragonal zirconia (YSZ, 95% ZrO_2, 5% Y_2O_3, *Tosoh/Nikkato*, Tokyo, Japan) were used in this research. The main properties of the balls are listed in Table 1.

Table 1. Properties of coating and ball materials.

Properties	AlCrN	YSZ
Hardness at 20 °C, GPa	30.6 ± 2.8 [5] *	10.5 **
Fracture toughness K_{IC}, MPa·m$^{0.5}$	6.4 [18]	6.0 **
Young's modulus of elasticity, GPa	585 ± 54 [5] *	210 **
Thermal conductivity 20 °C, W m^{-1}·K^{-1}	1.5 [19]	3 **
Max service temperature, °C	900 [20]	1200 **
Density, kg·m^{-3}	–	6000 **
Thermal diffusivity, ×10^{-6}·m^2·s^{-1}	–	0.9 [21]
Diameter, mm	–	10

Notes: * Property from coating from the same production line; ** Properties are provided by producer.

2.2. Characterization of Materials

Surface morphology was studied using a scanning electron microscope (SEM) *Hitachi TM1000* (Tokyo, Japan) equipped with an energy-dispersive X-ray source (EDS).

Raman spectral analysis was performed at room temperature using a high resolution micro-Raman spectrometer (*Horiba Jobin Yvon HR800*, Kyoto, Japan) equipped with a multichannel charge-coupled device (CCD) detection system. The device was set in the backscattering configuration. An Nd-YAG induced laser (λ = 532 nm) with a spot size of 10 μm in diameter was used for excitation. The results were obtained with laser beam powers of 2.8 mW (factor of filter was 0.0912) and 15.8 mW (factor of filter was 0.5147). The laser beam power was kept unchanged for all test materials (including transferred material). Crystal phases of the AlCrN PVD coating were detected by X-ray diffractometer (XRD) *Rigaku Ultima IV* (Tokyo, Japan). Obtained results were compared with measurement results from X-ray diffractometer *Bruker D5005 AXS* (Billerica, MA, US). A monochromatic CuKα radiation in 2θ scan mode was applied. Commercially available database ICDD-PDF-4+*2016* was used for the identifications of crystal phases.

2.3. Evaluation of Coating Reliability

Selected laboratory-scale experimental methods and approaches for improvement of reliability of the AlCrN PVD coating on the steel substrate for use in aggressive environments are listed in Table 2. The selection was done according to applicability for determination of coating reliability under static and tribologically influenced conditions. Visual observation, imaging techniques (optical, SEM, 2D or 3D profiling), ball cratering, adhesion or scratch testing are mainly suitable for preliminary estimation of properties and prediction of performance while electrochemical or tribo-corrosive tests (accompanied with electrochemical measurements) provide the possibility for tracking the performance of materials In-Situ. To great extent, only Raman spectroscopy can assist in an evaluation of the composition of a thin layer formed during tribological testing of material.

Table 2. Applicability of different methods for laboratory evaluation of various aspects of reliability of coatings intended for corrosive tribo-applications.

Task	Method												
	Visual and Tactile Observation	**Optical Microscopy**	**SEM**	**2D Contact Profilometry**	**3D Topography**	**KaloMax Ball Cratering**	**Adhesion Testing**	**Scratch Testing**	**Electrochemical Corrosion Tests**	**Tribo-Corrosive Test***	**EDS**	**Raman Spectroscopy**	**XRD**
Preliminary evaluation	+	+	+	++	+++	+++	+++	+++	+	NCH	++	++	+++
Wear or corrosion rate	-	+	+	+++	+++	NCH	NCH	NCH	++	+++	-	-	-
Destruction mechanisms	+	+	++	++	+++	NCH	NCH	+	+	++	+	+++	++
Elemental or phase composition of thin tribo-layer	-	-	- SE	NCH	NCH	NCH	NCH	-	0	0	-	+++	-
Elemental or phase composition of thick (≈>1 um) tribo-layer	-	0	+ BSE	NCH	NCH	NCH	NCH	-	0	0	++	+++	++
In-Situ measurement of corrosion intensity and/or evolution of coating damage	0	0	NCH	NCH	NCH	NCH	NCH	NCH	+++	+++	-	-	+

Notes: "+" is showing how useful could be the equipment (+ min, ++ average, +++ max); "-" means that equipment is rather not useful; "0"–only qualitative estimation or indirect conclusion; NCH–not considered here; SE and BSE–Secondary electron and Backscattered electrons of SEM; * Assisted with potentiostat/galvanostat.

2.3.1. Potentiodynamic Polarization Test

The typical three-electrode cell was used with the Pt counter-electrode (CE) with a working surface area of 2 cm^2 and the saturated calomel reference electrode (SCE). Potentiodynamic polarization measurements were performed in a 3.5 wt % NaCl aqueous solution to estimate the influence of the microdefects (micropores and macrodroplets) detected by SEM, Figure 2a. The corrosion current density (i_{corr}) was measured at room temperature to evaluation corrosion reaction kinetics. A corrosion potential ($E_{corr\ calc}$) was calculated from the intercept on Tafel plot. Pristine SS substrates, as well as coated substrates, were tested.

Specimens were isolated with a nonconductive tape and the remaining exposed surface area of about 1 cm^2 was used as working electrodes (WE). *Autolab PGSTAT30* galvanostat–potentiostat with general purpose electrochemical system (*GPES*) software (*Metrohm Autolab B.V., Utrecht*, The Netherlands) was employed for data recording. Open circuit potential (OCP) stabilization was done by immersing the samples into 3.5 wt % NaCl solution for either 10 min or 24 h before the test to estimate the change in polarization resistance. The limits of positive scanning linear sweep voltammetry were set from −0.7 up to 0 V for the coatings and from −0.8 up to 0 V for the substrate. Scanning rate was selected 5 $mV \cdot s^{-1}$. Software *NOVA* (version 2.1.2, *Metrohm Autolab B.V.*) was used to analyse the Tafel plot. Penetration rate *CR* (the thickness loss per unit of time $mm \cdot year^{-1}$]), protective efficiency P_i [%], porosity *F* [%] of the coating, the polarization resistances of the substrate and coating-substrate systems R_{pm} and R_p [$\Omega \cdot cm^{-2}$] were calculated according to ASTM G59-97e1-Standard Test Method for Conducting Potentiodynamic Polarization Resistance Measurements [22].

2.3.2. Tribo-Corrosion at Open Circuit Potential

Tribo-corrosion experiments were carried out using universal materials tester (UMT-2) from CETR (*Bruker*, Billerica, MA, US) in a reciprocating mode (amplitude 1×10^{-3} m, frequency 1 Hz). The counter ball was located above the specimen and, therefore, wear debris tend to remain in a wear scar. All tests were done in ambient atmosphere environment (temperature 20 ± 2 °C, relative humidity 50% ± 5%). The specimens were fixed in the electrochemical cell installed on the reciprocating table as shown in Figure 2b. The electrochemical cell was filled with 50 mL of 3.5 wt % NaCl aqueous electrolyte solution. A level of a liquid of 1 cm above the tribological contact was provided. The specimen was connected to the potentiostat as WE. Standard Ag/AgCl as RE and Pt as CE were utilized in the three-electrode mode. EmStat^{3+} potentiostat and *PSTrace* software (*PalmSens BV*, Houten, The Netherlands) was used for data recording and processing.

An exposed surface area of 1 cm^2 was left by isolating the remaining surface of specimens with a nonconductive tape. Exposed surfaces were cleaned step-by-step with acetone, ethanol and then dried before applying electrolyte. The material was immersed in the liquid for 1 h before test without data recording for preliminary stabilization. Recorded data for coated and uncoated materials were divided into three periods: (1) Stabilization (1000 s); (2) tribo-corrosion (7200 or 43,200 s that corresponds to 2 or 12 h); and (3) passivation (1000 s). The load was 1 kg (9.8 N) during 7200 or 43,200 s tests and 3 kg (29.4 N) during 7200 s tests. The initial maximum Hertzian contact pressure was either 1.31 or to 1.88 GPa for 1 or 3 kg tests, respectively.

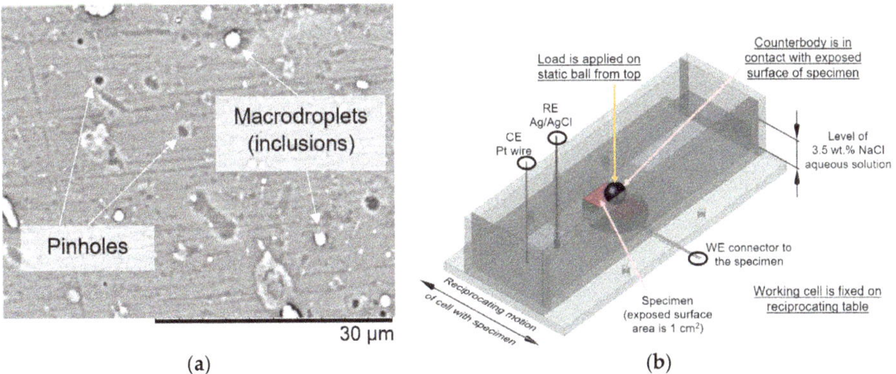

Figure 2. Surface defects on the as-deposited AlCrN coating on the stainless steel AISI 316L substrate were detected in a SEM micrograph (**a**); and a schematic illustration of reciprocating tribo-corrosion test setup (**b**).

3. Results and Discussion

3.1. Potentiodynamic Polarization Test of Statically Corroded Uncoated and Coated SS AISI 316L

The Tafel plots and calculated potentiodynamic polarization results of the SS substrate and the coated specimens are presented in Figure 3 and Table 3, respectively. The $E_{corr\ calc}$ and i_{corr} of the bare AISI 316L after 10 min of immersion were found to be −0.423 V vs. SCE and 1.9 μA·cm^{-2}, respectively. $E_{corr\ calc}$ shifts toward more positive value (to about −0.340 V vs. SCE) after applying the AlCrN PVD coating on the SS substrate. About 1.2 times lower i_{corr} was measured as compared to the AISI 316L, reaching improvement in the protective efficiency (P_i) by 15.9%.

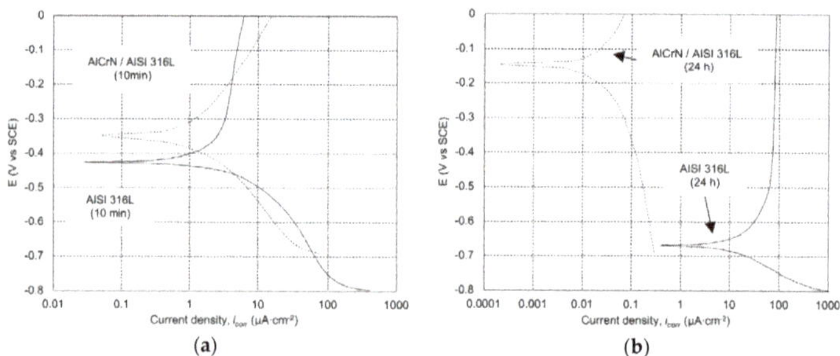

Figure 3. Potentiodynamic polarization curves (Tafel plots) of uncoated and AlCrN PVD coated AISI 316L specimens indicating corrosion potential: (**a**) after 10 min immersion; (**b**) after 24 h immersion.

Table 3. Potentiodynamic polarization of uncoated and coated specimens.

Material	Potentiodynamic Polarization Measurements (From NOVA)					Calculation Results		
	Corrosion Current Density	Polarization Resistance	Calculated Corrosion Potential	Tafel Slope	Tafel Slope	Corrosion Rate	Protective Efficiency	Porosity
	i_{corr}	R_{pm}, R_p	$E_{corr\ calc}$	$\lvert\beta_a\rvert$	$\lvert\beta_c\rvert$	CR	P_i	F
	[μA·cm^{-2}]	[Ω·cm^{-2}]	[V]	[V·Decade^{-1}]	[V·Decade^{-1}]	[mm·Year^{-1}]	[%]	[%]
AISI 316L, 10 min	1.9	1.43×10^4	−0.423	0.718	0.069	2.0×10^{-2}	–	–
AlCrN/AISI 316L, 10 min	1.6	3.76×10^4	−0.340	0.399	0.212	1.7×10^{-2}	15.9	29
AISI 316L, 24 h	22.1	1.71×10^3	−0.669	0.312	0.121	2.3×10^{-1}	–	–
AlCrN/AISI 316L, 24 h	0.05	2.33×10^6	−0.153	0.488	0.517	4.9×10^{-4}	99.8	0.002

A significant passivation of the AlCrN PVD coated sample was found after immersion for 24 h. $E_{corr\ calc}$ significantly shifts toward a positive value (−0.153 V vs. SCE) and i_{corr} was found to be ≈30 times lower (0.05 μA·cm^{-2}) as compared to the coating after 10 min immersion. The i_{corr} was also found being ≈40 and ≈450 times lower as compared with the tested substrate after 10 min and 24 h immersion, respectively.

Decrease in CR down to 4.9×10^{-4} mm·year^{-1} and F down to 0.002% for the AlCrN PVD coated sample was detected after 24 h immersion. It confirms a hydrolyzation reaction with the AlN on the surface and inside the pores. One of possible reaction is known as the Bowen's model [23] reporting degradation of the AlN in an aqueous medium. In the present work, the reaction passivates as the AlN is embedded into a monolithic dense coating and the reaction occurs only on the surface. Corrosion products such as an amorphous aluminium monohydroxide ($AlOOH_{(amorph)}$) and a crystalline aluminium hydroxide ($Al(OH)_{3(crystal)}$) form an additional passivation layer. Increased P_i up to 99.8% confirms the inertness of transition metal (e.g., Cr) nitrides [24].

3.2. Tribo-Corrosive Wear Test of Uncoated and Coated SS AISI 316L

Evolution in OCP before, during and after the short (2 h) wear tests of the bare and coated SS are presented in Figure 4a. Loss of protective oxide film developed on the surface of the bare SS was continuously observed during the whole stabilization period (1000 s) before starting the wear test. OCP shifts towards negative values from about −0.400 down to about −0.680 V vs. Ag/AgCl during stabilization.

Sharp shifting towards more negative values of OCP was measured at the beginning of wear test. Stabilized OCP value of about −0.780 V vs. Ag/AgCl remains practically unchanged until the end of the test. A sharp increase in coefficient of friction (CoF) from about 0.10 up to about 0.74 in the first 1200 s of the wear test (running-in period) indicates a rapid formation of a wear track representing a removed protective oxide layer and an enlarging tribocontact area. Instability in CoF was observed as a fluctuation from 0.70 to 0.74, exhibiting arrangement of the tribosystem (second running-in period). A slightly higher, but stable CoF of about 0.79 was measured after 2600 s of the test running, which indicates reaching a steady-state regime.

Some fluctuations in OCP from −0.230 to −0.160 V vs. Ag/AgCl were observed with the AlCrN PVD coated steel before wear test. These fluctuations are related to the solution penetration into the pores, other defects presented on the surface of the coatings, and reaction with the available AlN, causing a formation of the passive layer that consists of Al-based reaction products. Simultaneous evolution during the first 1600 s and stabilization in OCP and CoF values of about −0.210 V vs. Ag/AgCl and 0.58 were observed corresponding to the running-in and the steady-state regime, respectively. At these particular conditions, neither failure nor observable degradation of the coating was detected.

The test duration was increased up to 12 h with the application of 1 or 3 kg load as shown in Figure 4b. The fluctuations in OCP from −0.320 up to −0.250 V vs. Ag/AgCl were generally observed with the AlCrN PVD coated steel before wear test initiation, Figure 4b. Evolution of CoF during a running-in period (about 4000 s from the beginning of wear test) is similar to the evolution of CoF at 1 and 3 kg loads. An increase in CoF was measured from 0.42 up to the steady-state value of 0.69. A second CoF stabilization period was observed after about 21,000 s of test running, changing CoF to the final steady-state value of 0.65 (1 kg) and 0.62 (3 kg).

A passivation effect of the coated specimen was detected during the test under 1 kg load. The OCP slightly changed from −0.240 up to a more positive value of −0.150 V vs. Ag/AgCl. A rapid change to more negative OCP value of −0.300 V vs. Ag/AgCl was detected after 9000 s of sliding. It can be explained by an increased rate of a continuous mechanical destruction of the passivating layer. A slight passivation effect occurs after about 25,000 s as OCP was measured to be changing from −0.300 up to the final steady-state value of −0.280 V vs. Ag/AgCl.

Sharp shifting in OCP toward more negative values from −0.240 down to −0.440 V vs. Ag/AgCl was observed after the first 3300 s of the test under 3 kg load as shown in Figure 4b. This value remained stable for about 18,000 s of sliding indicating an inability of corrosion products to create a stable passivation layer under this load. However, the critical failure of the coating was recognized after about 18,000 s of the test run. The OCP was measured to be almost continuously shifting towards negative values from −0.440 down to about −0.700 V vs. Ag/AgCl at the end of the test.

It was also found that passivation effect of the immersed AlCrN coating occurs after about 45,400 s of static oxidation as demonstrated in Figure 5. The stabilized OCP is more positive than during tribo-corrosion tests (Figure 4b).

No typical iron based oxides on the pristine surface of the SS AISI 316L were found in the Raman spectra, Figure 6a. However, the Raman peak of Cr_2O_3 with a low intensity at 310 cm^{-1} indicates the development of the oxide layer at the ambient conditions. An increase in intensities of peaks collected from the area of a wear scar of the pristine SS points out to the formation of a thick layer of iron-based oxides and hydroxides [7]. A broad peak was observed in a range between 680 and 700 cm^{-1} indicates the presence of the corrosion product Fe_3O_4, Figure 6a. It should be noted that almost identical spectrum was obtained even after 24 h of static sample expose into the NaCl solution. Well-pronounced broad peaks appear in the spectra of the AlCrN PVD coated SS at 300, 690–706, 1000 and 1331–1388 cm^{-1}, Figure 6b. These peaks belong to the vibration of Cr and N ions and their intensities decrease after immersion in the 3.5 wt % NaCl solution for 12 h. The peak at 1000 cm^{-1} disappears, but a broad peak at 1331–1388 cm^{-1} turns into new peaks of low intensities as demonstrated in Figure 6b. It could be explained by the low and medium intensity combination of Raman active modes and overtones of α-Cr_2O_3 on the surface of coated material [25].

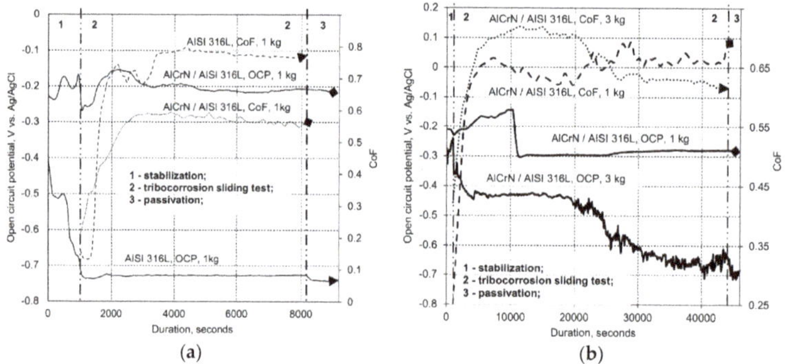

(a) (b)

Figure 4. Chronopotentiometry results to show comparison in evolution of CoF and open circuit potential (vs. Ag/AgCl) of uncoated and the AlCrN PVD coated AISI 316L samples during short (2 h) (**a**); and extended (12 h) tribo-corrosion test of coated AISI 316L sample (**b**).

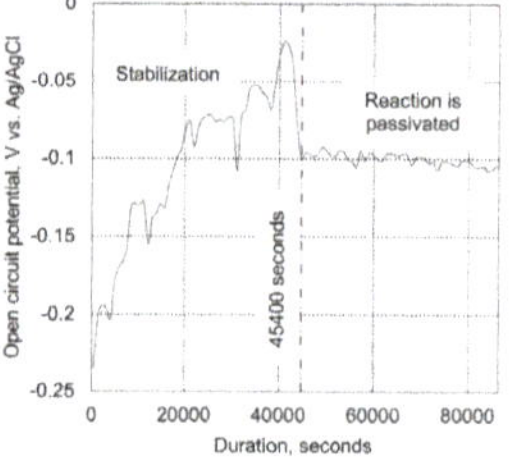

Figure 5. Chronopotentiometry result to show evolution of open circuit potential (vs. Ag/AgCl) of AlCrN PVD coated AISI 316L sample in the 3.5 wt % NaCl solution (static test duration–86,400 s/24 h).

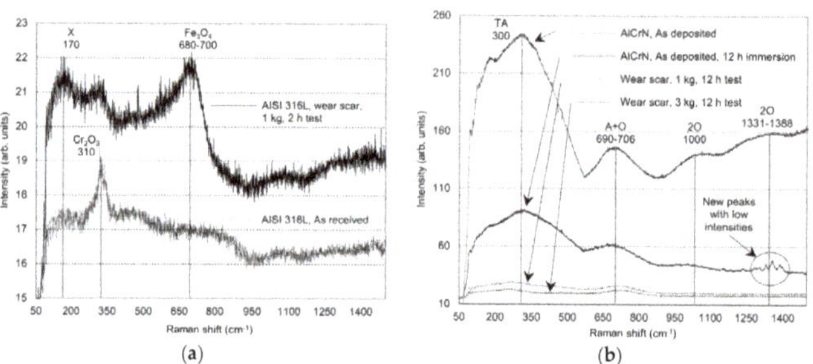

(a) (b)

Figure 6. Raman spectra illustrates vibration modes and oxides on pristine surface of AISI 316L and in a wear scar after 2 h (1 kg) tribo-corrosion wear test (**a**); and observed changes of intensities and shifting of peaks on coated specimens before and after 12 h immersion; and after extended (12 h) tribo-corrosion wear tests in 3.5 wt. NaCl % under 1 and 3 kg loads (**b**).

The typical grooves in the wear scar of the AISI 316L SS after the short tribo-corrosion test are demonstrated in Figure 7a. Several areas of the CrN rich interlayers and micro-droplet inclusions were exposed during an extended (12 h) tribo-corrosion test under 1 kg load, Figure 7b. Several areas of the uncovered CrN interlayer and extensive cracking in the middle region were found after the extended wear test under 3 kg load, Figure 7c. EDS analysis indicated a critical decrease in an atomic content of Al and appearance of elements typical for AISI 316L in the delaminated areas. The simplified schematic illustration of the static and dynamic corrosion processes is made based on the results obtained, Figure 8. The overview assignment of the Raman peaks of the uncoated and AlCrN coated SS AISI 316L before and after the corrosion and tribo-corrosion tests are presented in Table 4.

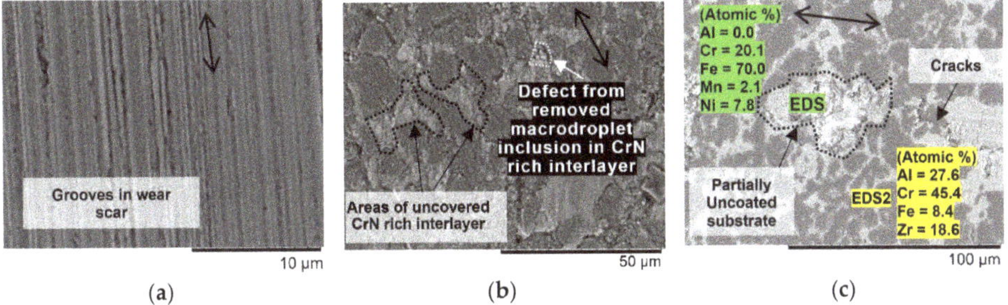

Figure 7. SEM micrographs show grooves inside wear scars after short (2 h) tribo-corrosion test with AISI 316L (**a**), and defects after extended (12 h) tribo-corrosion tests with AlCrN coated AISI 316L specimens under 1 kg (**b**), and 3 kg loads (**c**). Double arrowed line shows reciprocating sliding direction; elemental analysis was done with EDS inside and close to the marked area.

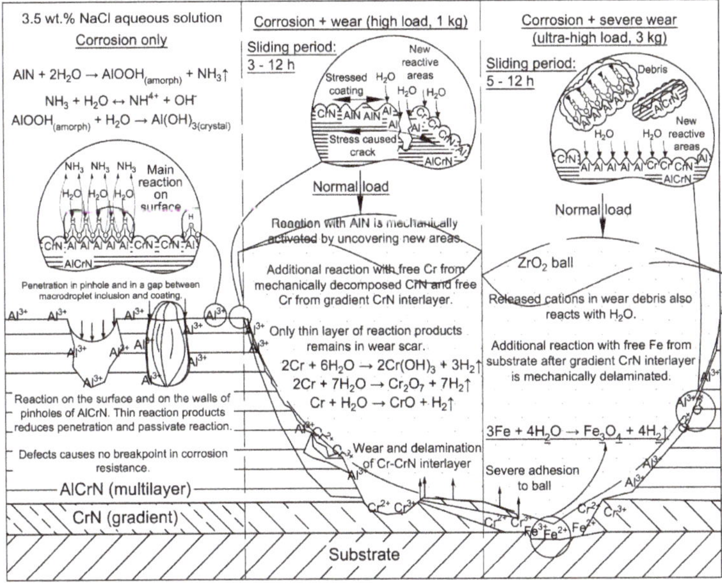

Figure 8. Simplified schematic illustration indicates main processes on the surface and in the wear scar during static corrosion and 12 h tribo-corrosion tests. A period when damage of interlayer or substrate took place in case of 1 or 3 kg tests, respectively is indicated. A ball is shown only for illustration without scaling. Mainly released cations are indicated.

Table 4. Assignment of Raman peaks before and after corrosion and tribo-corrosion tests.

Material	Test	Peak Position [cm^{-1}]	Peak Assignment	Peak Intensity	Peak Configuration	Comments
AISI 316L	As received	310	Cr_2O_3	Low	Sharp	Slight oxidation after polishing
	24 h static immersion	680–700	$Fe_3O_4/\gamma\text{-}Fe_2O_3$	Low	Broad	Development of Fe based oxides and hydroxides
	Tribo-corrosion 2 h, 1 kg	680–700	$Fe_3O_4/\gamma\text{-}Fe_2O_3$	Low	Broad	Development of Fe based oxides and hydroxides
AlCrN PVD on AISI 316L	As deposited	300	TA mode-vibration of Cr ions	High	Broad	Cubic CrN structure
		690–706	A+O optic mode-vibration of N ions	High	Broad	Cubic CrN structure
		1000	2 O-second order transition	Low	Broad	Cubic CrN structure
		1331–1388	2 O-second order transition	Low	Broad	Cubic CrN structure
	12 h static immersion	300	TA mode-vibration of Cr ions	High	Broad	*
		690–706	A+O optic mode-vibration of N ions	High	Broad	*
		1000	2 O-second order transition	Low	Broad	*
		1331–1388	-	Low	Sharp	Possible formation of Al based corrosion products
	Tribo-corrosion 12 h, 1 kg	300	TA mode-vibration of Cr ions	Low	Broad	*
		690–706	A+O optic mode-vibration of N ions	Low	Broad	*
	Tribo-corrosion 12 h, 3 kg	300	TA mode-vibration of Cr ions	Low	Broad	Formation of corrosion products
		690–706	A+O optic mode-vibration of N ions	Low	Broad	Formation of corrosion products.

Note: * Intensity decreases due to the formation of amorphous AlOOH [22].

4. Discussion

The reaction and the subsequent passivation effect in the 3.5 wt % NaCl solution can be attributed to the formation of a very thin and mainly amorphous layer on the surface of the AlCrN coating [26]. This layer consists mainly of $AlOOH_{(amorph)}$, which foremost is a result of the reaction between H_2O and AlN on the surface of the coating (including pinholes, gaps between microdroplet inclusions, etc.). The reactions are more intensive on the more defected areas of the surface. The applied incident powers of Nd-YAG induced laser (λ = 532 nm) of the Raman spectrometer checked from 0.05 up to 22 mW was not appropriate enough to detect this thin amorphous oxide layer due to massive side effects such as noise and/or weak signals. The CoF measurements in the conditions of the tribo-corrosive reciprocating sliding test demonstrate even lower final value at the ultra-high load (3 kg) as compared to the same test at a load of 1 kg. It indicates an intensive forming of a quite soft abrasive body of severely hydrolysed surfaces of the AlCrN-based wear debris (self-lubrication) at the extreme conditions as schematically shown in Figure 8. Self-lubrication effect provided by the hydrolyzation reaction with AlN can provide an improved reliability as a protective factor in a short period of overloading situations. It leads to significantly increased lifetime of coating in underwater conditions in addition to a high resistance to corrosion due to a presence of the interstitial compound of CrN.

It was found that the AlCrN coating is performing sufficiently better than TiCN or TiAlN coatings deposited onto the same SS substrate and tested by authors under the same tribo-corrosive conditions that indicate its higher reliability as the coating for protection of a soft steel substrate that can be produced by 3D printing (additive manufacturing technology) [2]. These TiCN and TiAlN coatings failed during 2-h sliding test with 1 kg load while the AlCrN coating was providing sufficient resistance up to the end of 12-h test with 1 kg load and failed only after 20,000 s of sliding with 3 kg load.

5. Conclusions

The multilayered AlCrN hard coating with the Cr adhesion and the gradient CrN interlayer was deposited on the SS substrate AISI 316L in a dominating cubic CrN structure by LARC PVD. The protective efficiency of the gradient AlCrN PVD coating increases up to 99.8% indicating a passivation layer developed through Al-based reaction products due to the reduced penetration rate of the corrosive media. The presence of surface defects (pinholes, inclusions, etc.) does not significantly affect the failure of coating in static corrosive conditions.

Tribo-corrosion tests performed in 3.5 wt % NaCl solution allows evaluation of the coating reactivity due to the AlN passivation effect combined with the presence of Al-based corrosion products. Change in OCP from −0.18 down to −0.3 V vs. Ag/AgCl after about 2 h-long sliding test under 1 kg load is related to the unprotected layer of Cr rich gradient CrN interlayer due to the partially lost layer of the AlCrN. An appearance and evolution of severe damages in the coating causes OCP shifting down to −0.42 V vs. Ag/AgCl at the beginning of sliding test under 3 kg load, indicating a mechanically initiated reaction with free Al from the AlN and free Cr from the AlCrN and CrN interlayer. The CoF of the coated samples remains about 25% lower (about 0.6) even after partial degradation as compared to the pristine SS under threefold lower load, indicating a reliability of the coating in tribo-contact even after a loss of chemical protection.

Author Contributions: Conceptualization, M.A. and I.H.; Methodology, J.B.; Software, T.R.; Validation, M.A., S.B. and I.H.; Formal Analysis, J.B.; Investigation, J.B.; Resources, M.A., S.B. and T.R.; Data Curation, J.B.; Writing-Original Draft Preparation, J.B.; Writing-Review & Editing, M.A. and I.H.; Visualization, J.B.; Supervision, M.A., I.H.; Project Administration, M.A.; Funding Acquisition, M.A. and I.H.

Funding: This research was supported by the Estonian Research Council under the personal grant (PUT1063) (I. Hussainova), Estonian Ministry of Higher Education and Research under Projects (IUT19-29 and IUT19-28), the European Union through the European Regional Development Fund, (Project TK141) and TTÜ base finance projects (B54, B56 and SS427) (M. Antonov).

Acknowledgments: The authors would like to thank Heinar Vagiström for the help with preparation of AlCrN PVD coatings. The authors would also like to thank Rainer Traksmaa and Arvo Mere for the help with XRD measurements.

Conflicts of Interest: The authors declare no conflict of interest. The founding sponsors had no role in the design of the study; in the collection, analyses, or interpretation of data; in the writing of the manuscript, and in the decision to publish the results.

References

1. Wood, R.J.K. Tribo-corrosion of coatings: A review. *J. Phys. D: Appl. Phys.* **2007**, *40*, 5502–5521. [CrossRef]
2. Holovenko, Y.; Antonov, M.; Kollo, L.; Hussainova, I. Friction studies of metal surfaces with various 3D printed patterns tested in dry sliding conditions. *Proc. Inst. Mech. Eng. Part J J. Eng. Tribol.* **2018**, *232*, 43–53. [CrossRef]
3. Ricker, R.E.; Papavinasam, S.; Berke, N.; Brossia, S.; Dean, S.W. On using laboratory measurements to predict corrosion service lives for engineering applications. *J. ASTM Int.* **2008**, *5*, 1–10. [CrossRef]
4. Fischer, A.; Mischler, S. Tribocorrosion: Fundamentals, materials and applications. *J. Phys. D Appl. Phys.* **2006**, *39*, 3120–3219. [CrossRef]
5. Antonov, M.; Afshari, H.; Baronins, J.; Adoberg, E.; Raadik, T.; Hussainova, I. The effect of temperature and sliding speed on friction and wear of Si_3N_4, Al_2O_3, and ZrO_2 balls tested against AlCrN PVD coating. *Tribol. Int.* **2018**, *118*, 500–514. [CrossRef]
6. Stack, M.M.; Antonov, M.M.; Hussainova, I. Some views on the erosion–corrosion response of bulk chromium carbide based cermets. *J. Phys. D Appl. Phys.* **2006**, *39*, 3165–3174. [CrossRef]
7. Larroumet, D.; Greenfield, D.; Akid, R.; Yarwood, J. Raman spectroscopic studies of the corrosion of model iron electrodes in sodium chloride solution. *J. Raman Spectrosc.* **2007**, *38*, 1577–1585. [CrossRef]

8. Fenker, M.; Balzer, M.; Kappl, H. Corrosion protection with hard coatings on steel: Past approaches and current research efforts. *Surf. Coat. Technol.* **2014**, *257*, 182–205. [CrossRef]
9. Dinu, M.; Mouele, E.; Parau, A.; Vladescu, A.; Petrik, L.; Braic, M. Enhancement of the corrosion resistance of 304 stainless steel by Cr–N and Cr(N,O) coatings. *Coatings* **2018**, *8*, 132. [CrossRef]
10. Gutsev, D.; Antonov, M.; Hussainova, I.; Grigoriev, A.Y. Effect of SiO_2 and PTFE additives on dry sliding of NiP electroless coating. *Tribol. Int.* **2013**, *65*, 295–302. [CrossRef]
11. Zhang, L.; Chen, Y.; Feng, Y.; Chen, S.; Wan, Q.; Zhu, J. Electrochemical characterization of AlTiN, AlCrN and AlCrSiWN coatings. *Int. J. Refract. Met. Hard Mater.* **2015**, *53*, 68–73. [CrossRef]
12. Baronins, J.; Podgursky, V.; Antonov, M.; Bereznev, S.; Hussainova, I. Electrochemical behaviour of TiCN and TiAlN gradient coatings prepared by lateral rotating cathode arc PVD technology. *Key Eng. Mater.* **2016**, *721*, 414–418. [CrossRef]
13. Kaindl, R.; Franz, R.; Soldan, J.; Reiter, A.; Polcik, P.; Mitterer, C.; Sartory, B.; Tessadri, R.; O'Sullivan, M. Structural investigations of aluminum-chromium-nitride hard coatings by Raman micro-spectroscopy. *Thin Solid Films* **2006**, *515*, 2197–2202. [CrossRef]
14. Johnson, D.F.; Jiang, D.E.; Carter, E.A. Structure, magnetism, and adhesion at Cr/Fe interfaces from density functional theory. *Surf. Sci.* **2007**, *601*, 699–705. [CrossRef]
15. Lukaszkowicz, K.; Sondor, J.; Balin, K.; Kubacki, J. Characteristics of CrAlSiN + DLC coating deposited by lateral rotating cathode arc PVD and PACVD process. *Appl. Surf. Sci.* **2014**, *312*, 126–133. [CrossRef]
16. Rana, D.; Mandal, B.M.; Bhattacharyya, S.N. Analogue calorimetric studies of blends of poly(vinyl ester)s and polyacrylates. *Macromolecules* **1996**, *29*, 1579–1583. [CrossRef]
17. Benedek, I.; Feldstein, M.M. Fundamentals of pressure sensitivity. In *Handbook of Pressure-Sensitive Adhesives and Products*; CRC Press: Boca Raton, FL, USA, 2008.
18. Kawate, M.; Kimura Hashimoto, A.; Suzuki, T. Oxidation resistance of $Cr_{1-x}Al_xN$ and $Ti_{1-x}Al_xN$ films. *Surf. Coat. Technol.* **2003**, *165*, 163–167. [CrossRef]
19. Samani, M.K.; Chen, G.C.K.; Ding, X.Z.; Zeng, X.T. Thermal conductivity of CrAlN and TiAlN coatings deposited by lateral rotating cathode arc. *Key Eng. Mater.* **2010**, *447–448*, 705–709. [CrossRef]
20. Nohava, J.; Dessarzin, P.; Karvankova, P.; Morstein, M. Characterization of tribological behavior and wear mechanisms of novel oxynitride PVD coatings designed for applications at high temperatures. *Tribol. Int.* **2015**, *81*, 231–239. [CrossRef]
21. Somiya, S.; Aldinger, F.; Spriggs, R.M.; Uchino, K.; Koumoto, K.; Kaneno, M. Handbook of Advanced Ceramics: Materials, Applications, Processing, and Properties. In *Handbook of Advanced Ceramics*; Elsevier: New York, NY, USA, 2003.
22. *ASTM G59-97e1 Standard Test Method for Conducting Potentiodynamic Polarization Resistance Measurements*; ASTM International: West Conshohocken, PA, USA, 1997.
23. Bowen, P.; Highfield, J.G.; Mocellin, A.; Ring, T.A. Degradation of aluminum nitride powder in an aqueous environmet. *J. Am. Ceram. Soc.* **1990**, *73*, 724–728. [CrossRef]
24. Yang, M.; Allen, A.J.; Nguyen, M.T.; Ralston, W.T.; MacLeod, M.J.; DiSalvo, F.J. Corrosion behavior of mesoporous transition metal nitrides. *J. Solid State Chem.* **2013**, *205*, 49–56. [CrossRef]
25. Gomes, A.S.O.; Yaghini, N.; Martinelli, A.; Ahlberg, E. A micro-Raman spectroscopic study of $Cr(OH)_3$ and Cr_2O_3 nanoparticles obtained by the hydrothermal method. *J. Raman Spectrosc.* **2017**, *48*, 1256–1263. [CrossRef]
26. Barshilia, H.C.; Selvakumar, N.; Deepthi, B.; Rajam, K.S. A comparative study of reactive direct current magnetron sputtered CrAlN and CrN coatings. *Surf. Coat. Technol.* **2006**, *201*, 2193–2201. [CrossRef]

Article

The Potential of Tribological Application of DLC/MoS$_2$ Coated Sealing Materials

Chao Wang [1,*], Andreas Hausberger [1], Philipp Nothdurft [2], Jürgen Markus Lackner [3] and Thomas Schwarz [4]

1. Polymer Competence Center Leoben GmbH, Roseggerstraße 12, 8700 Leoben, Austria; andreas.hausberger@pccl.at
2. Chair of Chemistry of Polymeric Materials, Montanuniversität Leoben, 8700 Leoben, Austria; philipp.nothdurft@unileoben.ac.at
3. Institute of Surface Technologies and Photonics, Joanneum Research Forschungsgesellschaft mbH, 8712 Niklasdorf, Austria; juergen.lackner@joanneum.at
4. SKF Sealing Solutions Austria GmbH, 8750 Judenburg, Austria; Thomas.Schwarz@skf.com

* Correspondence: chao.wang@pccl.at; Tel.: +43-3842-42962-85

Received: 13 June 2018; Accepted: 29 July 2018; Published: 31 July 2018

Abstract: The potential of the combination of hard and soft coating on elastomers was investigated. Diamond-like carbon (DLC), molybdenum disulfide (MoS_2) and composite coatings of these two materials with various DLC/MoS_2 ratios were deposited on four elastomeric substrates by means of the magnetron sputtering method. The microstructures, surface energy of the coatings, and substrates were characterized by scanning electron microscopy (SEM) and contact angle, respectively. The chemical composition was identified by X-ray Photoelectron Spectroscopy (XPS). A ball on disc configuration was used as the model test, which was performed under dry and lubricated conditions. Based on the results from the model tests, the best coating was selected for each substrate and subsequently verified in component-like test. There is not one coating that is optimal for all substrates. Many factors can affect the coatings performance. The topography and the rigidity of the substrates are the key factors. However, the adhesion between coatings and substrates, and also the coating processes, can impact significantly on the coatings performance.

Keywords: DLC; MoS_2; coating; elastomer; seals

1. Introduction

Coating is one of the approaches that can improve the tribological properties economically. In recent years, the development of the coating methods has opened up new possibilities to enhance the surface properties. Coatings can be generally divided into "soft coatings" and "hard coatings" [1]. Soft coatings, including soft metal (e.g., lead, indium) and lamellar solids (e.g., graphite and molybdenum disulfide (MoS_2)), provide good shearing characteristics and thus result in a reduction of friction. Hard coatings (e.g., diamond-like carbon (DLC), titanium nitride (TiN)) can improve protection against wear and present low wear rates.

The unique properties of elastomers, such as low modulus of elasticity, high Poisson's ratio, and high degree of resilience with low hysteresis, make elastomers very suitable for the application as seals. However, high and erratic friction under dry and starved lubrication conditions could increase the friction and wear rates. As a consequence of surface damage, the lifetime of seals can be shortened greatly [2]. An approach to reduce the friction under dry and starved lubrication conditions is to deposit DLC on rubber. A lot of studies, from deposition techniques to DLC composition on various rubber materials, such as nitrile butadiene rubber (NBR), hydrogenated nitrile butadiene rubber (HNBR), fluoroelastomer (FKM), and ethylene propylene diene monomer rubber (EPDM), has been done by a

Japanese group of Nakahigashi [3,4], Takikawa et al. [5–7], a Dutch group of Pei and Bui et al. [8–12], and other researchers [13,14]. MoS_2 as a solid lubricant is mostly employed with hard surfaces (e.g., metals, ceramics) [15–17]. As to the combination of the two coatings, Wang et al. [18] has deposited MoS_2 on Steels with a supporting DLC film and it showed the MoS_2/DLC compound film reduced the friction force in humid environment. Recently, Zhao et al. [19] has deposited the MoS_2/DLC multilayer coatings on Si wafer and steel in high humidity for aerospace industries and it showed a moderate improvement of the tribological properties. The influence of space irradiation on MoS_2/DLC composite film on Si and steel was investigated by Wu et al. [20]. It showed a reduction of the wear rate after irradiation, which could be related to the increase of hardness. Noshiro et al. [21] has studied the friction properties of sulfide/DLC coating with a nanocomposite or –layered structure on Si wafer, which shows better tribological properties than DLC film. Previous work has focused only on either the composite MoS_2/DLC coating on metals or DLC and MoS_2 separately as coating on elastomers. Therefore, more work is needed to investigate the potential of application of composite coatings on elastomers.

In this research, the tribological properties of DLC, MoS_2, and combined coatings of MoS_2 and DLC were investigated on four elastomers. Coated elastomers were tested in model tests and after that the results were verified in component-like tests. The influence factors of tribological behaviors are discussed. The aim of this study is to investigate the potential of tribological application of composite coatings of MoS_2 and DLC on elastomeric substrates for industrial seals, especially under starved lubrication conditions. In addition, the study provides a guideline to evaluate the coatings.

2. Experimental Details

2.1. Test Materials and Coatings

Four classical sealing materials were tested; i.e., fluoroelastomer (FKM), nitrile butadiene rubber (NBR), hydrogenated nitrile butadiene rubber (HNBR), and thermoplastic polyurethane (TPU). Among these four elastomers, FKM is the softest material, having a shore—a hardness of 84; followed by NBR (85) and HNBR (86). Due to its special chemical composition, TPU is the hardest material with a shore—A hardness of 95. For ball on disc tests, the samples were 20 mm × 20 mm square rubber sheets with a thickness of 2 mm, which were produced by the compression molding process. However, slight differences could be found on the surface under the microscope among TPU, HNBR, and FKM. For TPU a totally different molding die was used and the surface was polished. This is explained in more detail in Section 3.1.1 (microscopic analysis). For ring on disc tests, special samples were used, which are structurally similar to seals [22]. In order to remove contamination on the substrate and also inside the rubber (e.g., plasticizers [23]), all of the samples were cleaned using the standard cleaning procedures [9]. The difference between set and actual values can be explained with sputtering duration (Table 1). As a result of about three times longer sputtering duration time of the DLC 300 nm than the DLC 150 nm, the actual thickness of the DLC 300 nm is over three times thicker. The thickness of MoS_2 coating is proportional to the sputtering time. The thickness varied due to the influence of different sputtering parameters.

Table 2 shows the material and thickness of the investigated coatings. Five different materials (i.e., DLC, MoS_2, and three hybrid combinations of DLC and MoS_2 with various proportions) were deposited as coatings on the substrates. These two materials were not combined as multilayers, but rather in a composite. The proportion of MoS_2 in the composite increases from Hybrid_A to Hybrid_C. Based on our previous work, the set value of 300 nm was selected as the standard thickness for the coatings and the set values of the thickness were defined based on the deposition rate [24]. In order to investigate the influence of the thickness on the tribological properties, 150 nm thick coatings were also obtained through controlling the deposition process time. In order to measure the thickness, several samples were partially covered with tapes during the coating process. After removing the tapes, the thickness was measured with a contact stylus profilometry (Dektak 150 surface profiler, Veeco, Plainview, NY, USA).

In order to improve the adhesion of the coatings, prior to deposition a pre-treatment process was carried out by using a high vacuum experimentation bell jar system (Leybold Univex 450, Leybold Vacuum GmbH, Cologne, Germany) [25,26]. Substrates were fixed on the rotary table (φ = 560 mm) with a distance of 12 mm to the target.

The cylindrical pulsed laser deposition (PLD) evaporator was used as a target. The pre-treatment was performed at 3 kV DC acceleration voltage with 15 sccm Ar and 5 sccm O_2 gas flow. The chamber pressure was around 8.8×10^{-4} mbar. After pre-treatment, the coatings were deposited by means of the pulsed DC magnetron sputtering method. A graphite target (electrographite, 99.5% purity) was used as a sputtering source for DLC coatings. For MoS_2 its purity is 99.5%. Both targets were purchased from Sindlhauser Material GmbH (Kempten, Germany).

The parameters of the pre-treatment and deposition process are shown in Table 1. For pure DLC film, the ratio of C_2H_2/Ar was 0.19, due to the existence of C_2H_2, a-c: H film was generated [27,28]. For the hybrid coatings, only Ar was used as a source gas [28]. For the hybrid coatings, graphite and MoS_2 were ejected individually from two sputtering sources. Different hybrid variants were generated by varying sputter power. Remarkably, differences of the micro-structures can be observed on the coating when the substrates were deposited at different temperatures [29]. To avoid the thermal influences on substrates and coating processes, the pre-treatment and deposition processes were performed under constant ambient temperature (23 °C). However, due to plasma flow the temperature of the sample surface can increase up to 40 °C. After the deposition process, the samples were stored in Petri dishes in a box.

An optical microscope (Stereo Microscope SZX 12, Olympus, Tokyo, Japan) was employed to analyze the wear scars of the counterparts. The surface roughness was measured in three different regions of each sample with a three-dimensional focus variation microscope (InfiniteFocus, Alicona, Graz, Austria). The surface morphology and wear tracks of coated rubber were characterized with a scanning electron microscope (SEM, VEGA-II, TESCAN, Brno, Czech Republic).

In order to characterize the chemical composition of the coatings, X-ray photoelectron spectroscopy (XPS) analysis were carried out using a Thermo Scientific spectrometer with a micro-focused monochromatic Al Kα source (1486.6 eV, Thermo Fisher Scientific Inc., Waltham, MA, USA). All measurements were conducted with the radiation source operated at 12 kV and a beam current of 1.16 mA in a high vacuum below 10^{-7} mbar. A hemispherical analyzer was applied to detect the accelerated electrons. The electrons were collected from a spot area of 300 μm, which is vertical to the analyzer. To prevent charging and electron charge compensation of the samples, a flood gun was used.

Survey scans were acquired within an energy range of 0–1350 eV using a pass energy of 200 eV, a step size of 1.0 eV, a dwell time of 50 ms, and 2 scans. High resolution scans were obtained using a 50 eV pass energy, 0.1 eV step size, a dwell time of 50 ms, and 8 scans. For C 1*s*, Mo 3*d* and S 2*p*, binding energy ranges and total number of energy steps are as follows: 279–298 eV, 181 steps; 222–240 eV, 181 steps; 157–170 eV, 181 steps; respectively.

The spectra were referenced to the alkyl C 1*s* photoelectron peak at 284.8 eV, characteristic of the alkyl moieties (C–C/C–H). Peak positions for qualitative analysis are consistent with the corresponding assignment positions found in literature [30].

Spectra were analyzed using the Thermo Avantage software (Version 5932). The ratio of Lorentzian/Gaussian is 0.3. A standard Shirley background is used for the reference samples spectra. The spectra were fitted with Powell algorithm with a convergence of 10^{-6}. The maximum error for peak energy and full width at half maxima (FWHM) is ±0.1 eV. The sensitivity factors (SF) used for calculation are provided by the equipment supplier.

Table 1. Parameters of pre-treatment and deposition process.

Coating	Thickness (nm)	Pre-treatment		Deposition								
		Voltage (V)	Gas flow (sccm)	Sputtering source	Power (W)	Voltage (V)	Current (A)	Gas flow (sccm)	Pressure (mbar)	Rotation (rpm)	Duration (min)	Frequency (kHz)
DLC	300	3000	15 Ar + 5 O_2 for 5 min, 20 Ar for 25 min	Graphite	3000	577–578	5.21–5.22	42 Ar + 8 C_2H_2	2.3×10^{-3}	5.00	68	80
	150					579–582	5.22–5.19				23	
MoS_2	300			MoS_2	500	462–455	1.10–1.15	50 Ar	2.6×10^{-3}		60	
	150					468–461	1.10–1.13				30	
Hybrid_A	300			Graphite + MoS_2	C: 3000 MoS_2: 54	C: 602–601 MoS_2: 270–258	C: 4.95–4.93 MoS_2: 0.20–0.19	50 Ar	2.6×10^{-3}		65	
Hybrid_B	300				C: 3000 MoS_2: 255	C: 604–600 MoS_2: 402–403	C: 4.98–5.01 MoS_2: 0.64–0.66				54	
	150					C: 602–610 MoS_2: 405–404	C: 5.01–4.96 MoS_2: 0.65–0.67				27	
Hybrid_C	300				C: 3000 MoS_2: 440	C: 611–606 MoS_2: 467–446	C: 4.95–4.92 MoS_2: 0.98–1.04				36	

Table 2. Material and thickness of the coatings.

Material	Thickness (nm)		
	Set Value	Actual Value	Difference
DLC	300	405.0 ± 18.2	35.1%
DLC	150	113.3 ± 5.8	−24.1%
MoS_2	300	257.8 ± 19.2	−13.9%
MoS_2	150	131.8 ± 7.5	−12.2%
Hybrid_A	300	269.8 ± 14.0	−9.7%
Hybrid_B	300	300.2 ± 8.4	0.4%
	150	116.8 ± 6.0	−22.5%
Hybrid_C	300	246.3 ± 9.5	−17.8%

The determination of surface energy was carried out in a self-developed contact angle device. Distilled water and diiodomethane were applied as liquids to determine the polar and dispersive part of the surface energy, respectively. For each measurement, a drop of 2.5 μL volume was used. Each measurement was repeated three times. Owens et al. [31], Rabel [32] and Kaelble [33] method was applied for calculating the surface energy.

2.2. Test Procedures

The tribological properties were investigated by means of model tests and component-like tests. The model tests were performed on a micro tribometer with a ball on disc configuration (UMT-2, Bruker, Billerica, MA, USA). The development of the sample geometry for the component-like test was reported by Hausberger [22]. The tests were performed on a precision rotary tribometer (TE-93, Phoenix Tribology Ltd., Kingsclere, UK). Each test was repeated three times. All of the tests were conducted at room temperature (22 °C) with a relative humidity of 50% ± 10%. About one month after the coating process, the tribological tests were performed.

2.2.1. Ball on Disc Tests

Commercial 100Cr6 stainless steel balls of 6 mm diameter (HRC 60–62) were used as counterparts. The counter body slid on the elastomer at 100 mm/s with 1 N normal load. The radii of the run tracks were 5 mm and 7.5 mm. The total length of the tracks was 3.143×10^5 m. In order to obtain a better understanding of the function of the coatings, the tests were performed under dry and lubricated conditions. For the lubricated tests, approximately 7 mg Mobil SHC Grease 460WT (Viscosity of Oil, ASTM D 445 [31] cSt @ 40 °C = 460) was smeared equally over the whole surface [32]. The average thickness of the grease can be calculated. Its amount was chosen so that the thickness of the grease layer was approximately 0.02 mm.

2.2.2. Ring on Disc Tests

Ring-shaped counterparts of 34CrNiMo6 were used in the ring on disc test. They possessed an average roughness (R_a) of 0.035 μm. The ring-like sample was so constructed that there was only a line contact between the sample and counterpart [22]. The tests were conducted with 50 N normal load at room temperature (23 °C) and the speed of revolution was 118 rpm. The aim of this research is to improve the tribological properties of seals under starved lubricated conditions. In order to simulate starved lubrication condition in component-like tests, approximately 2 mg Mobil SHC Grease 460WT was smeared on the contact edge of the samples. For uncoated samples the tests lasted 168 h. For the coated samples, the tests were stopped automatically when the abort condition was reached. The abort condition was set according to the coefficient of friction of the uncoated samples. The principle of the ring on disc test is illustrated in Figure 1.

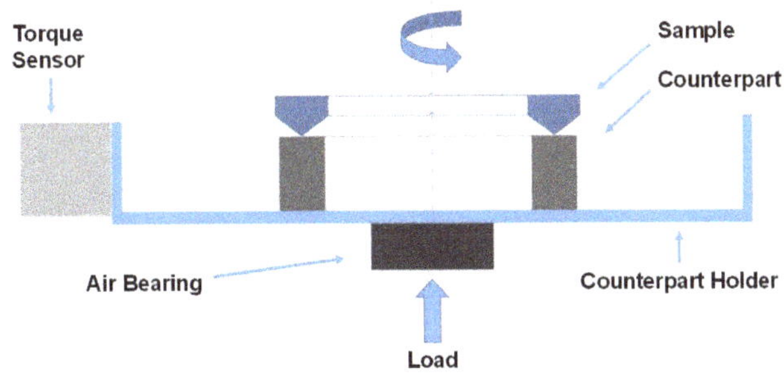

Figure 1. Principle of ring on disc test on TE-93.

The counterpart was fixed on the counterpart holder. The load, which was produced by a pneumatic pump, acted on the sample through the thrust bearing and counterpart. An electric motor was mounted on the top of the machine and drove the sample against the counterpart in a rotational movement. The torque, which was generated through friction, was measured by a torque sensor. Furthermore, the temperature near the contact area and in the middle of the counterpart was also measured during the test.

3. Results and Discussion

3.1. Characteristics of Coatings

After deposition the thickness of coatings was measured. The chemical composition was investigated with XPS measurements. The microstructures of the surfaces were analyzed with roughness and compared among different substrates. Furthermore, the surface energy of the substrates and coatings were identified.

3.1.1. Thickness of the Coatings

For each coating, the thickness was measured at six different positions of the two samples. Table 2 shows the set and actual average thickness. The difference between set and actual values can be explained with sputtering duration (Table 1). As a result of about three times longer sputtering duration time of the DLC 300 nm than the DLC 150 nm, the actual thickness of the DLC 300 nm is over three times thicker. The thickness of the MoS_2 coating is proportional to the sputtering time. The thickness varied due to the influence of different sputtering parameters.

As reported in [7], the application of C_2H_2 accelerates the deposition rate, which leads to a higher thickness than the set value.

3.1.2. Chemical Composition

The chemical composition, the assigned peak energies, full width at half maxima (FWHM), and sensitivity factor (SF) of each peak are given in Table 3 and were obtained with XPS analysis.

Table 3. Spectral fitting parameters.

Elements	Bonds	Peak Energy (eV)	FWHM (eV)	SF Al [34]	Ref.
C 1s	C–C/C–H	284.8	1.4	1.0	[33]
	C–O	286.0	2.1		[33]
	–COO	288.4	2.5		[33]
Mo 3*d*	MoS_2	229.0	2.0	5.6	[33,35]
	MoO_3	232.8	1.5		[33,35]
S 2*p*	S^{2-}	162.0	1.4	1.1	[33,36]
	${S_2}^{2-}$	163.6	1.4		[37,38]
S 2*s*	–	226.4	2.2	1.4	[38,39]

In Table 4, the chemical composition of each coating is listed. In order to avoid the influence of the different elastomeric substrates, coatings were deposited on silicon for the XPS analysis.

In both DLC coatings, the portion of C 1*s* is about 90% with no detectable silicon signal corresponding to a homogeneous carbon layer formation. The dominating carbon species are C–C/C–H bonds at 284.8 eV (Figure 2a) which are unambiguous assigned to the atomic structure of the used DLCs. The beneficial properties of DLC in tribology depend mainly on the similar hardness and Young's modulus as diamonds [28,40]. Besides, C–O and –COO signals were also found and are attributed to the surface oxidation during the coating process and storage [41] and is in good agreement with the results obtained in [42].

Table 4. Chemical compositions (C, O, Mo, S, and N) of the coatings.

Sample	Composition (%)							
	C	O	Mo		S	N	MoS_2/MoO_3	S/Mo
			MoS_2	MoO_3				
300 nm DLC	90.1	9.9	–	–	–	–	–	–
150 nm DLC	89.5	10.5	–	–	–	–	–	–
300 nm MoS_2	22.4	13.6	13.0	3.5	26.5	21.1	3.7	1.6
150 nm MoS_2	27.4	15.3	12.1	3.7	24.5	17.0	3.2	1.6
300 nm Hybrid_A	75.2	13.4	1.3	1.3	3.5	5.3	1.0	1.3
300 nm Hybrid_B	60.3	16.5	2.5	2.8	6.7	11.3	0.9	1.3
150 nm Hybrid_B	56.1	17.0	2.5	3.3	6.7	14.5	0.7	1.2
300 nm Hybrid_C	38.4	21.6	2.6	5.7	8.2	23.5	0.5	1.0

In pure MoS_2 coatings, the high nitrogen and carbon amounts are attributed to atmospheric contaminations (CO_2, hydrocarbons, N_2, etc.) during sample transport or storage or manufacturing of the samples. However, the S 2*p* doublet at 162.0 eV (ΔeV = 1.18) in combination with the doublet at 229.0 eV and 232.1 eV is unambiguous assigned to MoS_2 (Figure 1c,d). A second doublet in the Mo 3*d* spectra is attributed to Mo with environment as in MoO_3 [43]. On the subject of oxidation of molybdenum disulfide to molybdenum (VI) oxide, different reports were found [15,44]. In general, the oxidation rate is extremely low at ambient temperature and in the absence of a high concentration of moisture [15]. The oxidized layer at the outmost surface appears to protect the bulk material from further oxidation. However, different oxidation rates at ambient condition were investigated, and it was found the crystallite orientation plays an important role in the oxidation process [45]. Oxidation leads to a higher friction coefficient, enhanced wear rate, and hence a shorter wear life [36,46]. The ratio of MoS_2/MoO_3 in the reference coatings indicates that oxidation had occurred but the major portion is still MoS_2. For the hybrid samples (Figure 2d–f), the amount of MoS_2 is increased from Hybrid_A to Hybrid_C. As a result, the ratio of total S/Mo decreases from Hybrid_A (1.3) to Hybrid_C (1.0). The higher the concentration of MoS_2, the higher the oxidation rate and as a result the lower the MoS_2/MoO_3 ratio. It is suggested that the increase of MoS_2 amount in the coatings accelerates the oxidation rate and is highest for the Hybrid_C sample.

The S 2*p* spectrum appears as two overlapping doublets. This means, different types of sulfur ligands, such as bridging terminal $S_2{}^{2-}$, and bridging S^{2-} species exist in the coating [37]. In addition, our results were in agreement with Benoist et al. observations as a higher oxygen content lead to a decrease in the S^{2-} sulphur component whereas the $S_2{}^{2-}$ pair increase [43].

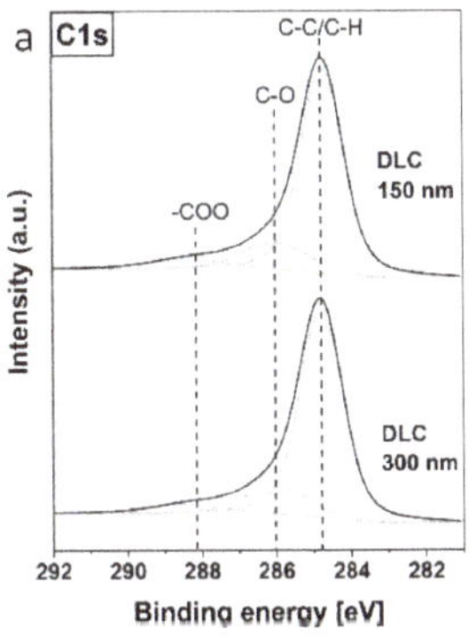

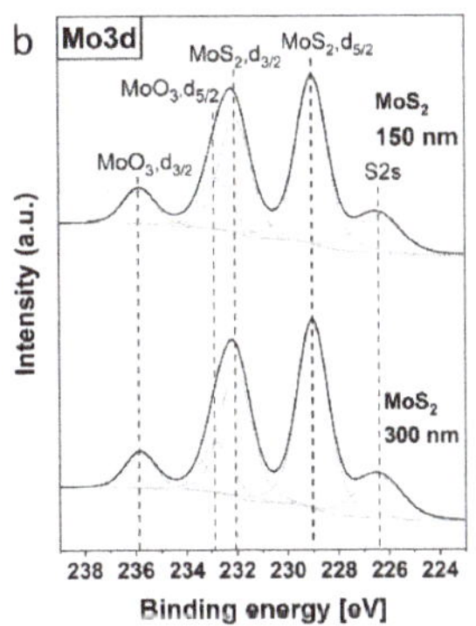

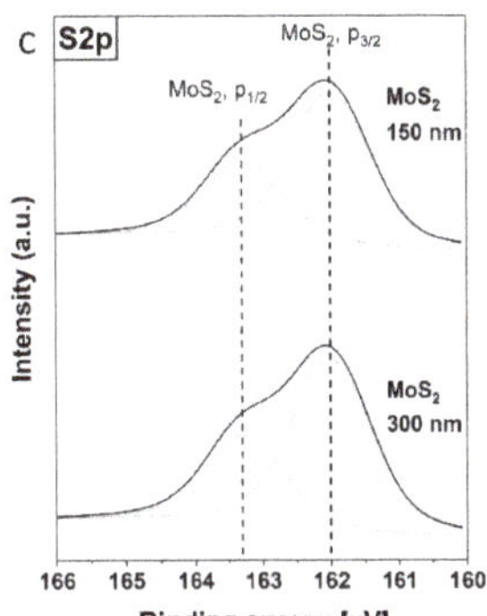

Figure 2. *Cont.*

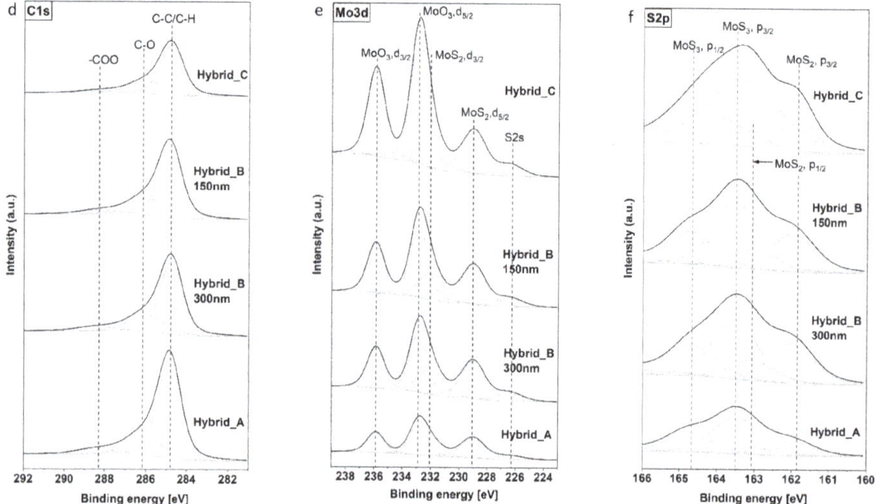

Figure 2. X-ray photoelectron spectroscopy (XPS) spectra: C 1s of diamond-like carbon (DLC) coatings (**a**) and hybrid coatings (**d**); Mo 3*d* of MoS_2 coatings (**b**) and hybrid coatings (**e**); S 2*p* of MoS_2 coatings (**c**) and hybrid coatings (**f**).

3.1.3. Microscopic Analysis

The uncoated substrates, except HNBR, were analyzed with a microscope and presented in a previous work [24]. Generally, on a macroscopic level the uncoated HNBR, NBR, and FKM possess similar parallel, strip-like structures, whereas TPU presents completely different structures. Due to the different physical properties of elastomers, especially elasticity and viscosity, which can have an effect on the flow properties in the molding process, they behaved differently during the processing [47]. Although uncoated HNBR and FKM show similar macrostructures, on the microscopic level utterly different microstructures can be observed. The surface of uncoated HNBR is relatively smooth but with some small debris. However, the surface of uncoated FKM is much rougher and with dense particles. This can also be explained with R_a and R_z. In spite of the very similar R_a value of uncoated HNBR and FKM, the R_z value of uncoated FKM is more than 30% higher than that of uncoated HNBR (Table 5).

Table 5. The average roughness (R_a), mean roughness depth (R_z) of uncoated samples.

Parameter	FKM	HNBR	NBR	TPU
R_a (μm)	1.00	1.03	0.61	0.44
R_z (μm)	6.74	5.59	3.69	3.98

The surface of uncoated TPU was full of small strips. However, the strips were not as neatly arranged as those of HNBR and FKM. Also, compared to HNBR and FKM, the strips on TPU were much narrower and shallower. Another difference, which must be mentioned, is that, except for the strips, there were almost no small debris or particles on uncoated TPU.

In this section, for each substrate two coatings have been chosen and discussed. The two coatings were so chosen that, regarding the substrate, one of them showed the best tribological performance and the other showed the worst in the dry and lubricated ball on disc tests. In addition, one thin coating for HNBR was chosen to analyze the influence of the thickness.

The strip-like structures on the surface, which can be observed in Figure 3a, were produced because of the compression molding process. Not only can these structures be found in HNBR, but also in FKM and NBR. Figure 3c shows one other position from the same sample as Figure 3a. Not like the rough surface in Figure 3b, flake-like structures with small debris can be observed in Figure 3d. Moreover, cracks can be observed on the surface. As reported by Takikawa and Pei, cracks are typical surface structures of DLC coated rubber [5,12,29].

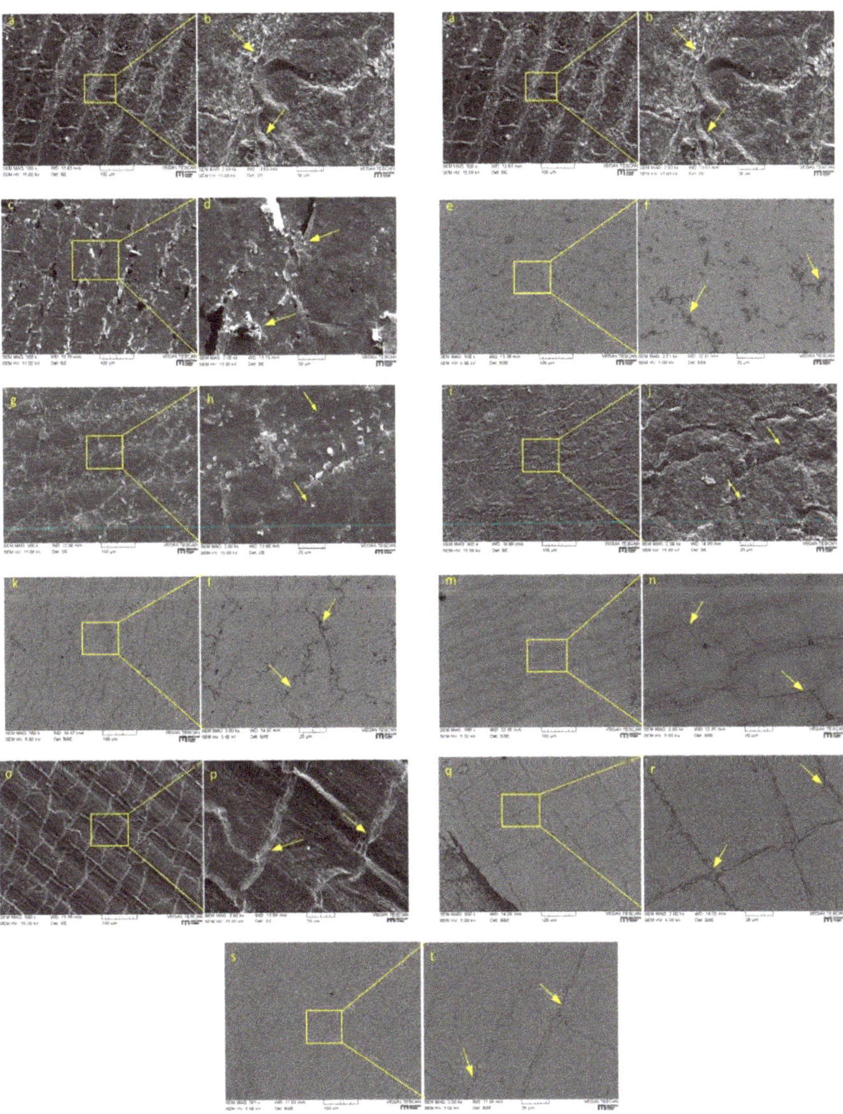

Figure 3. With 300 nm (**a**,**c**) and 150 nm DLC (**e**) coated hydrogenated nitrile butadiene rubber (HNBR); with 300 nm Hybrid_A coated HNBR (**g**), fluoroelastomer (FKM), (**i**) and nitrile butadiene rubber (NBR) (**o**); with 300 nm MoS_2 coated FKM (**k**), NBR (**m**) and thermoplastic polyurethane (TPU) (**s**); with 300 nm DLC coated TPU (**q**). High magnification (**b**,**d**,**f**,**h**,**j**,**n**,**p**,**r**,**s**,**t**) are shown to the right side of the respective low magnification (500×).

Compared with the 300 nm DLC coating, the 150 nm DLC coating looks smoother on the whole. However, small particulates can be observed on the surface (Figure 3f). Figure 3g shows the microstructures of 300 nm Hybrid_A on HNBR. Scaly microstructures were observed and they look similar to the DLC coating to some degree (Figure 3b). As previously mentioned, Hybrid_A is a composite coating, which possesses the least MoS_2 among the three hybrid coatings. However, a small amount of MoS_2 changed the microstructures considerably. The gaps between each piece of debris are smaller and the coating is noticeably smoother. This can be attributed to the much lower hardness of MoS_2 compared to DLC [15,48]. It seems that MoS_2 lowered the average hardness. Therefore, the coating can be better suited to the substrates roughness.

Comparing Figure 3h,j shows that the coating roughness was influenced to some extent by the substrate properties. Moreover, under high magnification, small holes can be identified on FKM with a 300 nm Hybrid_A coating (Figure 3j). That means the coating did not totally adhere to the substrate. This can be caused by the lower wettability of FKM compared to HNBR (details in Section 3.1.5). Small holes can also be observed on MoS_2 coated FKM (Figure 3k). However, the coating from MoS_2 looks much finer and smoother than the hybrid coating.

Generally, the surfaces of coated NBR are smoother compared to coated HNBR. Also, it should be emphasized that almost no debris could be found on the surface after coating. Moreover, as can be observed in Figure 3o, cracks which were caused by the removal of the sample from the deposition chamber, are rather neatly arranged on the surface, either parallel or perpendicular to the original microstructure of the substrate.

For TPU samples, they do not have the strip-like, neatly arranged microstructures like other substrates (Figure 3q,s). Because of its shallower and sparser microstructures, the roughness of the TPU substrate is correspondingly lower. Like the previous comparison, MoS_2 coated TPU is also finer and smoother than the DLC coated TPU (Figure 3s).

From the above comparisons, several influence factors that contribute to the coating microstructures were found and discussed. Firstly, the substrate topography is one of the most important influence factors for the coating microstructure. That is because of the smaller thickness (150–300 nm) compared to the roughness of the substrate (Table 5). Secondly, the composition of the coating plays an important role as well. Generally, on DLC or DLC-included coating small debris can be observed. In comparison with DLC, the MoS_2 coating is finer and smoother. Thirdly, the coating microstructures can be influenced by the material properties of the substrates in several ways. Coatings on a substrate like FKM, which has a lower wettability, show a higher possibility that the coating becomes porous and loose. Thermal properties (e.g., thermal expansion coefficient and thermal conductivity), are also influence factors. As shown in Table 6, FKM expands the most among the four materials, when the temperature increases by a given degree. This can lead to the scaly coating, which can be observed in Figure 3i,k.

Table 6. Thermal parameters of used materials.

Parameter	FKM	HNBR	NBR	TPU
Coefficient of thermal expansion (10^{-6}/K)	191	166	165	160
Thermal conductivity (W/(m·K))	0.24	0.15	0.26	0.06

However, the importance of these factors depends to a large extent on the ambient conditions of the coating process. In addition, the microstructure of the coatings is also affected by the deposition condition. For DLC coating, a-c: H was produced by using a plasma of Ar and C_2H_2, while for other coatings, only Ar was used. Generally, the coatings prepared with C_2H_2 look smoother than those without C_2H_2. This is in good agreement with the results in [7].

3.1.4. Surface Roughness

The surface roughness of the coated samples was affected by the substrate surface and also the microstructures of coatings, which can be changed by removing the samples from the deposition chamber (Figure 3o). In addition, the surface microstructure can also affect the adherence of the coating [49]. Generally, a rougher surface can have a better adhesion with coating because more bonding connections can be created. However, the scale of dimensions of the surface microstructures must be less than the film thickness [50]. As can be seen from Figure 4, uncoated substrates have different degrees of roughness. FKM and HNBR possess a similar roughness ($R_a \approx 1.0$ μm), while TPU and NBR have an appreciably lower roughness value. As mentioned previously, two different molding dies were used to produce samples, one for FKM, NBR, and HNBR, the other one for TPU. Moreover, the surface microstructure can be affected by additives, which could come up on the surface.

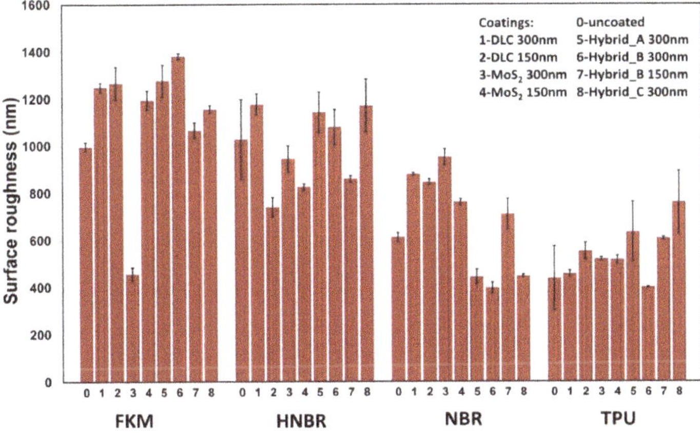

Figure 4. Surface roughness (R_a) of uncoated and coated substrates.

As can be seen from Figure 3a,e, the HNBR with 150 nm DLC is smoother than that with 300 nm DLC. However, for the other three materials with DLC coating, the thickness does not play an important role in the surface roughness. Compared to the uncoated FKM, the roughness of 300 nm MoS_2 coated FKM was reduced drastically, whereas the roughness of DLC-containing coatings on FKM was increased to varying degrees. This could be attributed to the larger difference between DLC and substrates in hardness and brittleness [48]. The DLC coatings with a very low thickness could be broken into fractures easily, when the coated samples are removed from the deposition chamber with a small deformation. However, this phenomenon was not found on other materials. For hard material TPU, no obvious differences could be identified in roughness. On the one hand, due to its different processing, its surface is smoother than other materials. On the other hand, higher hardness prevents its deformation by removal.

3.1.5. Surface Energy

One of the conditions for good wetting is that the surface tension of the substrate is higher than that of the still liquid coating material [51]. To eliminate the influences of substrates, coatings were also deposited on silicon. As shown in Figure 5, uncoated Si and coated Si possess higher surface energies than the other four substrates. Surface roughness plays an important role for the surface energy [52]. For four elastomeric substrates, surface energies were increased to varying degrees after coating. On the one hand, through comparison of the microstructures before and after deposition, it can be found that it changed significantly. Although the mean roughness (R_a) of the substrates had not been changed in a very large way, the microstructures were totally modified after the coating process. This leads to a modification of the surface energy. On the other hand, from the perspective of the material, the surface energy of elastomer substrates [53], DLC and MoS_2 are also different. These two factors together affect the difference of the surface energy after coating.

Compared with an uncoated elastomer, silicon shows a much higher surface energy in both polar and dispersed parts. After coating silicon shows a similar surface energy to the elastomers. Generally, FKM has almost the lowest surface energy in all coatings. Except for the influence of its chemical structure, the surface microstructure of uncoated FKM is different from HNBR and NBR. Comparing to HNBR, which has a similar mean roughness (R_a), FKM is much rougher and dense particles can be seen on the surface [24]. The film thickness has only a limited effect on surface energy.

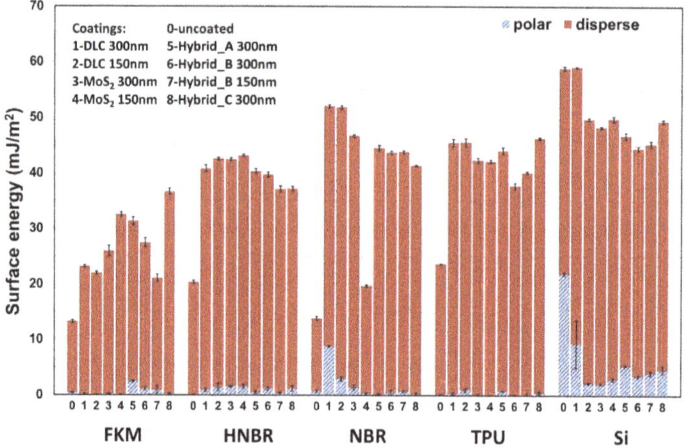

Figure 5. Surface energy of uncoated and coated substrates.

3.2. Tribological Tests

In order to study the potential of DLC/MoS_2 coatings on elastomers for tribological applications, the coatings were firstly tested in the model test under dry and lubricated conditions, so that the coatings could be evaluated comprehensively. Subsequently, the best and worst coatings were selected and investigated under starved lubrication condition in component-like tests.

3.2.1. Coefficient of Friction

The coefficients of friction (COF) for uncoated and coated elastomers in dry and lubricated ball on disc sliding tests are shown in Figure 6. Under dry sliding ambient conditions, almost all of the coatings bring an advantage to the tribological properties. In particular for HNBR, with 300 nm DLC coating, the COF was reduced from 0.99 to 0.18 by 82%. For NBR the frictional reduction, which the 300 nm DLC or Hybrid_A coating brought, was also significant; approximately 74%. For FKM and TPU, the decrease was not so appreciable. What was interesting was that for TPU the COF was slightly brought down by 300 nm Hybrid_B coating. The values of these measurements are in good agreement with the values reported in the literature [3,12,54]. However, when the thickness of the coating was reduced to 150 nm, the COF increased by 11%, compared to the uncoated TPU. This can be explained by microscopic analysis. As can be observed in Figure 7, the 150 nm coating was already severely damaged (Figure 7b) and the elastomeric substrate had direct contact with the counterpart during the test, while the thick variant was still intact (Figure 7a). That means for this coating, the thickness plays an essential role with respect to the tribological properties. However, the thickness cannot bring a significant difference in every case. That depends on several factors, for example, the hardness of the substrate, the coating material, and the adherence of coating material on the counterpart, the microstructures of the surface, the adherence between coating and substrate, lubrication conditions and so on. Adhesion and deformation are the two most important mechanisms that are responsible for the frictional behaviors of elastomers [55,56]. The high friction of uncoated HNBR and NBR under dry conditions show that not only deformation, which can be related to the relatively low hardness, but also adhesion, which can be seen as a dissipative stick-slip process on molecular level, are influential factors for the dry frictional behaviors [57–60]. This is in good agreement with Rabinowicz's studies, which indicated that low ratios of surface energy/hardness are associated with better surface interactions and also less adhesion [61,62]. Moreover, because of the high friction more dynamic energy would be expected to be transformed into heat, which could lead to an increase of temperature. Based on this conjecture the material's hardness will reduce with a higher temperature so that it could experience a higher wear rate [63].

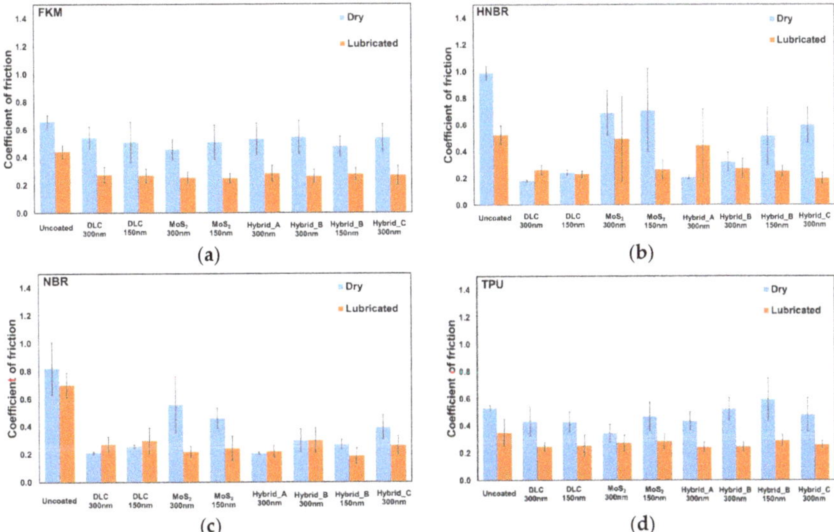

Figure 6. Coefficient of friction of uncoated and coated elastomers in dry and lubricated ball on disc sliding tests. (**a**) FKM; (**b**) HNBR; (**c**) NBR; (**d**) TPU.

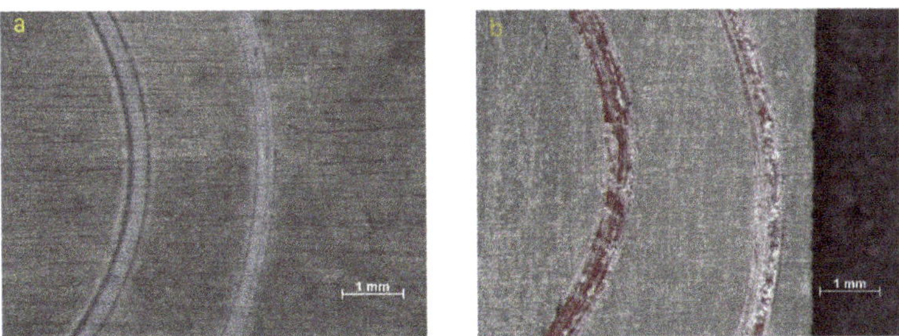

Figure 7. Microscopic images of wear tracks: (**a**) 300 nm Hybrid_B coated TPU; (**b**) 150 nm Hybrid_B coated TPU.

Under lubricated conditions, the differences of COF among various coatings were not as evident as in dry tests. The lubricant has no significant impact on the COF. One reason for this is that lubricants facilitate the stick-slip process on the molecular level to some extent. Therefore, the adhesion part for friction can be decreased [64]. As to the deformation part, it was assumed to stay on a similar level as under dry conditions in two aspects. One aspect is that the lubricant film is very thin on the contact area hence the stiffness of the film is negligible. The other aspect is that the COF of various coatings is similar.

It should be noted that in some cases the lubricant even brought a slight, negative impact on the tribological properties for DLC coated HNBR and NBR. This can be explained with two main reasons. One important aspect is that because of the high viscosity of the grease used, more energy would be needed to overcome the fluid friction [64,65]. The COF under dry conditions was extremely low. In this case, the benefit of the lubricant was less than its disadvantage. That means more energy was needed to overcome the resistance, which was brought by the lubricant.

Based on the results of the ball on disc tests under dry and lubricated conditions, the best coatings were chosen and verified in the component-like test (ring on disc). For FKM with Hybrid_A coating, its dry COF is slightly lower (2.5%) than Hybrid_B 300 nm. However, its lubricated COF is about 13% higher than Hybrid_B 300 nm. For HNBR with Hybrid_A coating, it is clear that MoS_2 brings a negative effect for the tribological performance. In the dry tests, the coatings with pure MoS_2 or high content of MoS_2 (Hybrid_B and Hybrid_C) were broken after the tests. Therefore, these coatings were not taken into consideration for the selection. In addition, as references, uncoated substrate and the worst coatings were also tested. The best and worst coatings are presented in Table 7. As can be seen from this table, the soft coating MoS_2 provides the best tribological properties for the softest material FKM, whereas the hard coating DLC is the best choice for the hardest material HNBR, only among FKM, NBR, and HNBR. For NBR, which has a middle hardness, a hybrid coating is better than other coatings. Because of its totally different surface structures, TPU was not comparable with the other elastomers.

Table 7. The best and worst coating for each material from ball on disc tests.

Material	Best Coating	Worst Coating
FKM	300 nm MoS_2	300 nm Hybrid_A
NBR	300 nm Hybrid_A	300 nm MoS_2
HNBR	300 nm DLC	300 nm Hybrid_A
TPU	300 nm MoS_2	300 nm DLC

For the ring on disc tests an abort condition was set up so that when the coating was worn or damaged, the test would be stopped immediately. As abort condition, an average COF of the uncoated substrate under stable running conditions was employed. As shown in Figure 8, at the beginning of the tests, for HNBR and TPU the COF of the best and worst coatings were almost at the same level. However, the COF of the uncoated substrate kept at a constant level after the running-in phase with a higher value, while the COF of the worst coating started to increase gradually. After just several hours, the friction was raised to the same level as the uncoated substrate. Compared to the worst coating, the best variant lasted significantly longer until the COF reached the abort condition. This means that the coating failed with increasing test time. Therefore, for HNBR, NBR, and TPU the trends of validation show a good correspondence with the results from the ball on disc tests. However, for FKM with the best coating, after the loading phase, its COF was already slightly over the abort condition, which represents the COF of uncoated FKM. It was found that the coating was already damaged. This implies that the combination of soft coatings like MoS_2 and soft substrate like FKM is inappropriate for this line contact. Because of its low hardness, FKM showed a strong local deformation under line contact. According to Archard's wear law for adhesive wear [66], wear volume is inversely proportional to the hardness of a substrate. By this situation, in which the contact area is relatively small, the soft coating on a soft substrate could be worn quickly.

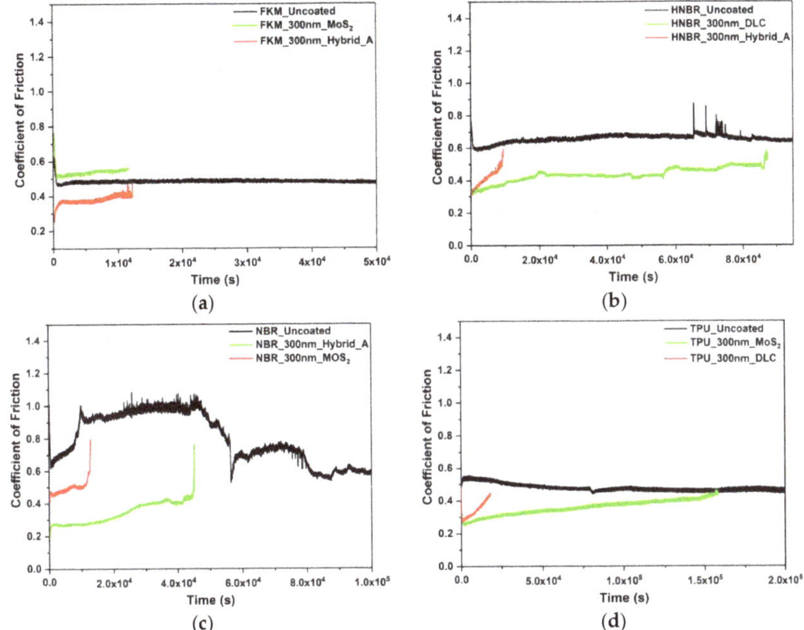

Figure 8. Comparison of the COF: uncoated and coated in Ring on disc test. (**a**) FKM; (**b**) HNBR; (**c**) NBR; (**d**) TPU.

3.2.2. Wear

The SEM images (Figure 9) show the wear track of the 300 nm DLC coated HNBR after a dry ball on disc test. It is evident that in the majority of the run area, the surface got smoother. The DLC coating was slightly pressed down due to the normal load and the microstructures were plastically deformed because of the tangential traction, which was generated by the sliding motion. Some piles of small crystal-like fragments can be found on the run track (Figure 10g). DLC is a very hard material and the thickness of the coating is just 300 nm. That means that when the counter body slid over the surface, both the elastomeric substrate and the coating experienced a deformation. The difference is that the substrate deformed viscoelastically and the coating showed a plastic deformation. Meanwhile, the cracks of the coating can also be ascribed to the enormous difference in hardness between the two materials.

Two positions of the wear track of 300 nm Hybrid_A coated HNBR were shown in Figure 9c,e. Particles can be observed in the troughs, which were located between every two peaks. White particles (Figure 3a,f) can be MoO_3, the oxidation product of MoS_2, which has a negative effect on the performance [15,36]. As shown in Table 4, in hybrid coatings, MoO_3 possesses larger portion than MoS_2. According to [67], when less than 30% of the MoS_2 converted to MoO_3, wear performance is still good. However, when it is greater than 50%, the wear behavior gets poor. As can be seen from Figure 3a,g, a part of the particles were generated during the coating process. Particles were also generated through dynamic motion in crack area. All of these particles were collected during the test in the trough. As can be observed in Figure 9d, some of the particles were pressed on the surface when the ball slid over.

From the same coating and substrate, sheet-like wear particles are visible in Figure 9e,f. This phenomenon can be attributed to surface fatigue [68]. Due to the repeated plastic deformation, sheet-like particles were gradually generated and separated from the coating.

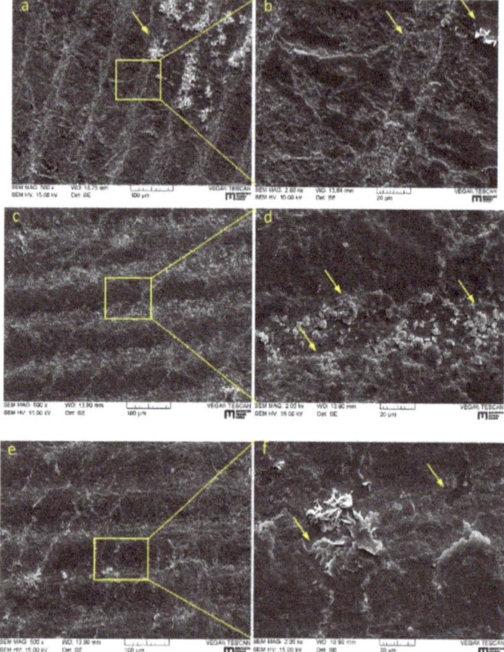

Figure 9. SEM micrographs: wear track of 300 nm DLC coated HNBR (**a**) and 300 nm Hybrid_A coated HNBR (**c**,**e**). Related areas are marked and shown with high magnification (**b**,**d**,**f**).

Compared to the 300 nm Hybrid_A coating on FKM before (Figure 3i) and after (Figure 10a) the test, a great number of cracks was generated during the test. This can be related to the dense particle-like microstructures of uncoated FKM. When the porous and loose coating was pressed by the counter body, it deformed more heavily and easily than other coatings. Besides, due to its lowest hardness among the four elastomers, the deformation of FKM is the largest. These two reasons could explain this phenomenon.

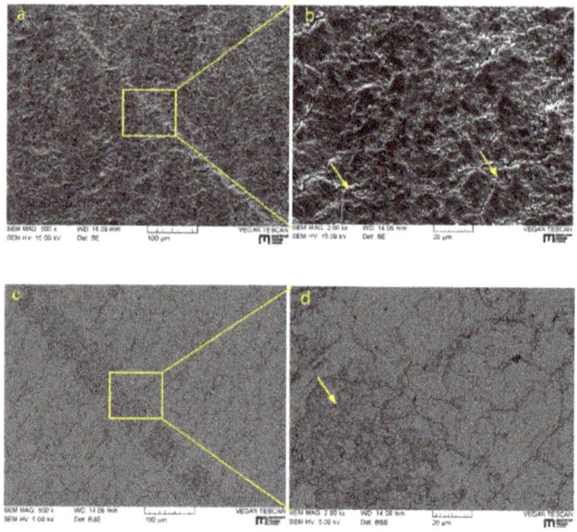

Figure 10. *Cont.*

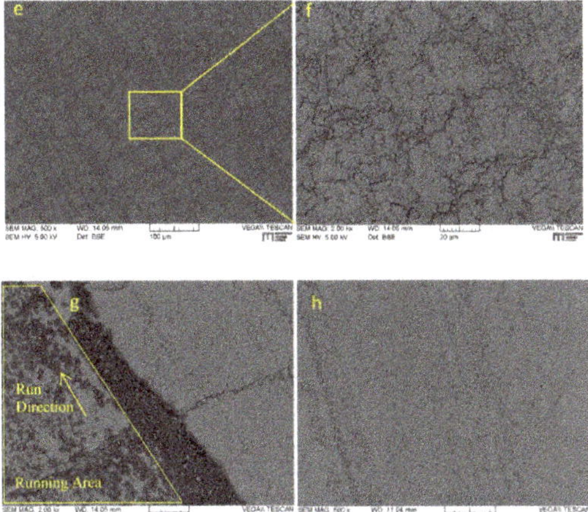

Figure 10. Scanning electron microscope (SEM) micrographs: wear track of 300 nm Hybrid_A coated FKM (**a**), 300 nm MoS_2 coated FKM (**c**), 300 nm DLC coated TPU (**e**,**g**), and 300 nm MoS_2 coated TPU (**h**). Related areas are marked and shown with high magnification (**b**,**d**,**f**).

Not like the 300 nm Hybrid_A coating, no obvious alteration could be found on the 300 nm MoS_2 coating after 10,000 cycles. Only the contact area was pressed and subsequently crushed into small pieces (Figure 10c). This can be attributed to the S–Mo–S sandwich structure of MoS_2, which facilitates the sliding motion on its surface [15].

The wear track of 300 nm DLC coated TPU (Figure 10e,f) presented very similar microstructures as MoS_2 coated FKM. That means only the DLC coating in the contact area was pressed into small pieces. However, plenty of wear particles, which are around 1 μm, were found close to the edge of the run track (Figure 10g). In some areas, they were piled up together. At the beginning of the test, the DLC coating was pressed into small pieces. However, some of the small particles that were detached from the substrate, rolled down from the sides to the middle of the groove. More and more particles were gathered on the lane with more cycles. At this moment, the particles were pushed out of the lane when the counter body slid over. Still quite a number of particles were found on the track after the test. Apparently, the dynamic movement of these small particles has influenced the tribological behavior to some extent. This can explain why DLC is the best coating for TPU under lubricated conditions but presented worse tribological properties than MoS_2 in dry tests. There is a strong possibility that under lubricated conditions the wear particles can be carried out of the track by grease. This is also one of the main functions of a lubricant [69].

Because of its low hardness and good shear characteristics no obvious particles were found on the MoS_2 coated TPU. Slight abrasive wear can be observed on the surface (Figure 10h). This is also one of the major wear processes on polymers [70]. Due to its special properties and good adherence on TPU, 300 nm MoS_2 shows the best tribological properties in dry tests.

4. Conclusions

The concept of the combination of hard and soft coatings on elastomers has been investigated. In this research, DLC was taken as an example of a hard coating and MoS_2 as a soft coating. It was proven that this concept can be used to improve the tribological properties of elastomers, especially under starved lubrication condition. There is not one coating that is optimal for all substrates. For different rubber substrates, the coating should be chosen individually, based on the substrate, coating properties, and their interaction. For a rubber substrate with low rigidity like FKM, soft coatings like MoS_2 present better tribological properties than hard coatings like DLC. This is attributed to the good shear characteristics and good deformation properties of MoS_2. Meanwhile, for a substrate with a higher rigidity like HNBR, a hard coating like DLC is a better option. For NBR, whose rigidity is between FKM and HNBR, a hybrid coating is the best choice. It possesses both advantages of hard and soft coatings. For TPU, due to its totally different microstructures, a different wear mechanism was discussed. For a hard substrate with a smooth surface, MoS_2 presented a better performance than a hard coating because the small particles of the hard coating can bring disadvantages during sliding motions.

Through the observation of microstructures on uncoated and coated surfaces the influence of the surface roughness and surface energy on tribological properties was investigated. The low surface energy of substrate leads to a porous and loose coating. As a consequence, the tribological properties could be adversely influenced.

The concept of the combination of hard and soft coatings will open new fields for the use of coatings in tribological applications on elastomers. Our data rule out the possibility that the application of DLC/MoS_2 as a coating can improve the tribological properties of elastomeric seals, especially under dry or insufficiently lubricated conditions. This finding is promising and should be explored with different combinations of even more than two coatings.

Author Contributions: Conceptualization, A.H., J.M.L., and T.S.; Methodology, A.H., J.M.L., and C.W.; Validation, C.W.; Formal Analysis, C.W., A.H., P.N., T.S.; Investigation, C.W.; Data Curation, C.W.; Writing–Original Draft Preparation, C.W., P.N.; Writing–Review & Editing, A.H., J.M.L., T.S.; Visualization, C.W.; Data Supervisor, A.H., T.S.; Resources, T.S., J.M.L.; Project Administration, A.H., J.M.L., T.S.; Funding Acquisition, T.S.

Funding: This research was funded by the project of "Bionics4Efficiency", which is one project of the "Bridge Program" (84037) of the Austrian Research Promotion Agency (FFG).

Acknowledgments: The authors gratefully thank M. Mitterhuber, W. Waldhauser and H. Parizek for their technical and scientific support and useful discussions.

Conflicts of Interest: The authors declare no conflict of interest.

References

1. Holmberg, K.; Matthews, A. *Coatings Tribology. Properties, Mechanisms, Techniques and Applications in Surface Engineering*, 2nd ed.; Elsevier Science: Amsterdam, The Netherlands, 2009.
2. Gawliński, M. Friction and wear of elastomer seals. *Arch. Civ. Mech. Eng.* **2007**, *7*, 57–67. [CrossRef]
3. Nakahigashi, T.; Tanaka, Y.; Miyake, K.; Oohara, H. Properties of flexible DLC film deposited by amplitude-modulated RF P-CVD. *Tribol. Int.* **2004**, *37*, 907–912. [CrossRef]
4. Nakahigashi, T.; Miyake, K.; Murkami, Y. Application of DLC coating to rubber and polymer materials. *J. Jpn. Soc. Tribol.* **2002**, *47*, 833–839.
5. Takikawa, H.; Miyakawa, N.; Minamisawa, S.; Sakakibara, T. Fabrication of diamond-like carbon film on rubber by T-shape filtered-arc-deposition under the influence of various ambient gases. *Thin Solid Films* **2004**, *457*, 143–150. [CrossRef]
6. Takikawa, H.; Miyakawa, N.; Toshifuji, J.; Minamisawa, S.; Matsushita, T.; Takemura, K.; Sakakibara, T. Preparation of elastic DLC film on rubber by T-shape filtered arc deposition. *IEEJ Trans. Fundam. Mater.* **2003**, *123*, 738–743. [CrossRef]
7. Miyakawa, N.; Minamisawa, S.; Takikawa, H.; Sakakibara, T. Physical–chemical hybrid deposition of DLC film on rubber by T-shape filtered-arc-deposition. *Vacuum* **2004**, *73*, 611–617. [CrossRef]
8. Bui, X.; Pei, Y.; de Hosson, J.T.M. Magnetron reactively sputtered Ti-DLC coatings on HNBR rubber: The influence of substrate bias. *Surf. Coat. Technol.* **2008**, *202*, 4939–4944. [CrossRef]
9. Bui, X.L.; Pei, Y.T.; Mulder, E.D.G.; de Hosson, J.T.M. Adhesion improvement of hydrogenated diamond-like carbon thin films by pre-deposition plasma treatment of rubber substrate. *Surf. Coat. Technol.* **2009**, *203*, 1964–1970. [CrossRef]
10. Pei, Y.; Bui, X.; de Hosson, J.T.M. Deposition and characterization of hydrogenated diamond-like carbon thin films on rubber seals. *Thin Solid Films* **2010**, *518*, S42–S45. [CrossRef]

11. Pei, Y.; Martinez-Martinez, D.; van der Pal, J.P.; Bui, X.; Zhou, X.; de Hosson, J.T.M. Flexible diamond-like carbon films on rubber: Friction and the effect of viscoelastic deformation of rubber substrates. *Acta Mater.* **2012**, *60*, 7216–7225. [CrossRef]
12. Pei, Y.; Bui, X.; Zhou, X.; de Hosson, J.T.M. Tribological behavior of W-DLC coated rubber seals. *Surf. Coat. Technol.* **2008**, *202*, 1869–1875. [CrossRef]
13. Lackner, J.M.; Major, R.; Major, L.; Schöberl, T.; Waldhauser, W. RF deposition of soft hydrogenated amorphous carbon coatings for adhesive interfaces on highly elastic polymer materials. *Surf. Coat. Technol.* **2009**, *203*, 2243–2248. [CrossRef]
14. Kahn, M.; Menegazzo, N.; Mizaikoff, B.; Berghauser, R.; Lackner, J.M.; Hufnagel, D.; Waldhauser, W. Properties of DLC and Nitrogen-Doped DLC Films Deposited by DC Magnetron Sputtering. *Plasma Process. Polym.* **2007**, *4*, S200–S204. [CrossRef]
15. Lansdown, A.R. *Molybdenum Disulphide Lubrication*, 1st ed.; Elsevier Science: Amsterdam, The Netherlands, 1999.
16. Bellido-González, V.; Jones, A.H.S.; Hampshire, J.; Allen, T.J.; Witts, J.; Teer, D.G.; Ma, K.J.; Upton, D. Tribological behaviour of high performance MoS_2 coatings produced by magnetron sputtering. *Surf. Coat. Technol.* **1997**, *97*, 687–693. [CrossRef]
17. Donnet, C.; Martin, J.M.; Le Mogne, T.; Belin, M. Super-low friction of MoS_2 coatings in various environments. *Tribol. Int.* **1996**, *29*, 123–128. [CrossRef]
18. Wang, D.; Chang, C.; Ho, W. Microstructure analysis of MoS_2 deposited on diamond-like carbon films for wear improvement. *Surf. Coat. Technol.* **1999**, *111*, 123–127. [CrossRef]
19. Zhao, X.; Lu, Z.; Wu, G.; Zhang, G.; Wang, L.; Xue, Q. Preparation and properties of DLC/MoS_2 multilayer coatings for high humidity tribology. *Mater. Res. Express* **2016**, *3*, 066401. [CrossRef]
20. Wu, Y.; Liu, Y.; Yu, S.; Zhou, B.; Tang, B.; Li, H.; Chen, J. Influences of space irradiations on the structure and properties of MoS_2/DLC lubricant film. *Tribol. Lett.* **2016**, *64*, 24. [CrossRef]
21. Noshiro, J.; Watanabe, S.; Sakurai, T.; Miyake, S. Friction properties of co-sputtered sulfide/DLC solid lubricating films. *Surf. Coat. Technol.* **2006**, *200*, 5849–5854. [CrossRef]
22. Hausberger, A.; Godor, V.; Grün, F.; Pinter, G.; Schwarz, T. Development of ring on disc tests for elastomeric sealing materials. In Proceedings of the International Tribology Conference, Tokyo, Japan, 16–20 September 2015.
23. Rodgers, B.; Waddell, W. The science of rubber compounding. In *Science and Technology of Rubber*, 4th ed.; Mark, J.E., Erman, B., Roland, C.M., Eds.; Academic Press: Cambridge, MA, USA, 2013; pp. 417–471.
24. Thirumalai, S.; Hausberger, A.; Lackner, J.M.; Waldhauser, W.; Schwarz, T. Effect of the type of elastomeric substrate on the microstructural, surface and tribological characteristics of diamond-like carbon (DLC) coatings. *Surf. Coat. Technol.* **2016**, *302*, 244–254. [CrossRef]
25. Lackner, J.M.; Waldhauser, W.; Schwarz, M.; Mahoney, L.; Major, L.; Major, B. Polymer pre-treatment by linear anode layer source plasma for adhesion improvement of sputtered TiN coatings. *Vacuum* **2008**, *83*, 302–307. [CrossRef]
26. Assender, H.; Bliznyuk, V.; Porfyrakis, K. How surface topography relates to materials' properties. *Science* **2002**, *297*, 973–976. [CrossRef] [PubMed]
27. Cho, N.-H.; Krishnan, K.M.; Veirs, D.K.; Rubin, M.D.; Hopper, C.B.; Bhushan, B.; Bogy, D.B. Chemical structure and physical properties of diamond-like amorphous carbon films prepared by magnetron sputtering. *J. Mater. Res.* **1990**, *5*, 2543–2554. [CrossRef]
28. Robertson, J. Diamond-like amorphous carbon. *Mater. Sci. Eng. R Rep.* **2002**, *37*, 129–281. [CrossRef]
29. Martinez-Martinez, D.; de Hosson, J.T.M. On the deposition and properties of DLC protective coatings on elastomers: A critical review. *Surf. Coat. Technol.* **2014**, *258*, 677–690. [CrossRef]
30. Moulder, J.F. *Handbook of X-ray Photoelectron Spectroscopy. A Reference Book of Standard Spectra for Identification and Interpretation of XPS Data*; Perkin-Elmer: Eden Prairie, MN, USA, 1992.
31. *ASTM D445-17a Test Method for Kinematic Viscosity of Transparent and Opaque Liquids (and Calculation of Dynamic Viscosity)*; ASTM International: West Conshohocken, PA, USA, 2017.
32. *Data Sheet of Mobil SHC™ Grease 460 WT*; Exxon Mobil Corporation: Irving, TX, USA, 2016.
33. *NIST X-ray Photoelectron Spectroscopy Database*; Version 4.1; National Institute of Standards and Technology: Gaithersburg, MD, USA. Available online: https://srdata.nist.gov/xps/Default.aspx (accessed on 13 June 2018).

34. Scofield, J.H. Hartree-slater subshell photoionization cross-sections at 1254 and 1487 eV. *J. Electron Spectrosc. Relat. Phenom.* **1976**, *8*, 129–137. [CrossRef]
35. Xu, Y.; Zheng, C.; Wang, S.; Hou, Y. 3D arrays of molybdenum sulphide nanosheets on Mo meshes: Efficient electrocatalysts for hydrogen evolution reaction. *Electrochim. Acta* **2015**, *174*, 653–659. [CrossRef]
36. Fleischauer, P.D.; Lince, J.R. A comparison of oxidation and oxygen substitution in MoS_2 solid film lubricants. *Tribol. Int.* **1999**, *32*, 627–636. [CrossRef]
37. Weber, T.; Muijsers, J.C.; Niemantsverdriet, J.W. Structure of Amorphous MoS_3. *J. Phys. Chem.* **1995**, *99*, 9194–9200. [CrossRef]
38. Baker, M.A.; Gilmore, R.; Lenardi, C.; Gissler, W. XPS investigation of preferential sputtering of S from MoS_2 and determination of MoS_x stoichiometry from Mo and S peak positions. *Appl. Surf. Sci.* **1999**, *150*, 255–262. [CrossRef]
39. Qiu, L.; Xu, G. Peak overlaps and corresponding solutions in the X-ray photoelectron spectroscopic study of hydrodesulfurization catalysts. *Appl. Surf. Sci.* **2010**, *256*, 3413–3417. [CrossRef]
40. Robertson, J. Classification of diamond-like carbons. In *Tribology of Diamond-Like Carbon Films: Fundamentals and Applications*; Donnet, C., Ed.; Springer: Boston, MA, USA, 2008; pp. 13–24.
41. Paik, N. Raman and XPS studies of DLC films prepared by a magnetron sputter-type negative ion source. *Surf. Coat. Technol.* **2005**, *200*, 2170–2174. [CrossRef]
42. Leung, T.Y.; Man, W.F.; Lim, P.K.; Chan, W.C.; Gaspari, F.; Zukotynski, S. Determination of the sp^3/sp^2 ratio of a-C: H by XPS and XAES. *J. Non-Cryst. Solids* **1999**, *254*, 156–160. [CrossRef]
43. Benoist, L.; Gonbeau, D.; Pfister-Guillouzo, G.; Schmidt, E.; Meunier, G.; Levasseur, A. XPS analysis of lithium intercalation in thin films of molybdenum oxysulphides. *Surf. Interface Anal.* **1994**, *22*, 206–210. [CrossRef]
44. Martincová, J.; Otyepka, M.; Lazar, P. Is single layer MoS_2 stable in the air? *Chemistry* **2017**, *23*, 13233–13239. [CrossRef] [PubMed]
45. Fleischauer, P.D. Effects of crystallite orientation on environmental stability and lubrication properties of sputtered MoS_2 thin films. *ASLE Trans.* **2008**, *27*, 82–88. [CrossRef]
46. Fleischauer, P.D.; Bauer, R. Chemical and structural effects on the lubrication properties of sputtered MoS_2 films. *Tribol. Trans.* **1988**, *31*, 239–250. [CrossRef]
47. Theilade, U.A.; Hansen, H.N. Surface microstructure replication in injection molding. *Int. J. Adv. Manuf. Technol.* **2007**, *33*, 157–166. [CrossRef]
48. Donnet, C.; Erdemir, A. *Tribology of Diamond-Like Carbon Films. Fundamentals and Applications*; Springer: Boston, MA, USA, 2008.
49. Neuville, S.; Matthews, A. A perspective on the optimisation of hard carbon and related coatings for engineering applications. *Thin Solid Films* **2007**, *515*, 6619–6653. [CrossRef]
50. Baglin, J.E.E. Interface design for thin film adhesion. In *Fundamentals of Adhesion*; Lee, L.-H., Ed.; Springer: Boston, MA, USA, 2014; pp. 363–382.
51. Goldschmidt, A.; Streitberger, H.-J. *BASF Handbook on Basics of Coating Technology*, 2nd ed.; Vincentz Network: Hannover, Germany, 2007.
52. Quéré, D. Wetting and Roughness. *Annu. Rev. Mater. Res.* **2008**, *38*, 71–99. [CrossRef]
53. Martínez, L.; Nevshupa, R.; Álvarez, L.; Huttel, Y.; Méndez, J.; Román, E.; Mozas, E.; Valdés, J.R.; Jimenez, M.A.; Gachon, Y.; et al. Application of diamond-like carbon coatings to elastomers frictional surfaces. *Tribol. Int.* **2009**, *42*, 584–590. [CrossRef]
54. Van der Pal, J.P.; Martinez-Martinez, D.; Pei, Y.T.; Rudolf, P.; de Hosson, J.T.M. Microstructure and tribological performance of diamond-like carbon films deposited on hydrogenated rubber. *Thin Solid Films* **2012**, *524*, 218–223. [CrossRef]
55. Grosch, K.A. The relation between the friction and visco-elastic properties of rubber. *Proc. R. Soc. A Math. Phys. Eng. Sci.* **1963**, *274*, 21–39. [CrossRef]
56. Zhang, S.-W. *Tribology of Elastomers*, 1st ed.; Elsevier: Amsterdam, The Netherlands, 2004.
57. Fuller, K.N.G.; Tabor, D. The effect of surface roughness on the adhesion of elastic solids. *Proc. R. Soc. A* **1975**, *345*, 327–342. [CrossRef]
58. Persson, B.N.J. On the theory of rubber friction. *Surf. Sci.* **1998**, *401*, 445–454. [CrossRef]
59. Mofidi, M.; Prakash, B. Influence of counterface topography on sliding friction and wear of some elastomers under dry sliding conditions. *Proc. Inst. Mech. Eng. Part J J. Eng. Tribol.* **2008**, *222*, 667–673. [CrossRef]

60. Glaeser, W.A.; Brundle, C.R.; Evans, C.A. *Characterization of Tribological Materials*, 2nd ed.; Momentum Press: New York, NY, USA, 2012.
61. Rabinowicz, E. Influence of surface energy on friction and wear phenomena. *J. Appl. Phys.* **1961**, *32*, 1440–1444. [CrossRef]
62. Rabinowicz, E. *Friction and Wear of Materials*, 2nd ed.; Wiley: New York, NY, USA, 1995.
63. Friedrich, K. *Friction and Wear of Polymer Composites*; North Holland Publishing Co.: Amsterdam, The Netherlands, 1986.
64. Bowden, F.P.; Tabor, D. *The Friction and Lubrication of Solids*; Oxford University Press: London, UK, 1963.
65. Okrent, E.H. The effect of lubricant viscosity and composition on engine friction and bearing wear. *ASLE Trans.* **1961**, *4*, 97–108. [CrossRef]
66. Archard, J.F. Contact and rubbing of flat surfaces. *J. Appl. Phys.* **1953**, *24*, 981–988. [CrossRef]
67. Lince, J.R.; Frantz, P.P. Anisotropic oxidation of MoS_2 crystallites studied by angle-resolved X-ray photoelectron spectroscopy. *Tribol. Lett.* **2001**, *9*, 211–218. [CrossRef]
68. Booser, E.R. *CRC Handbook of Lubrication. (Theory and Practice of Tribology). Theory and Design*; CRC Press: Boca Raton, FL, USA, 1983.
69. Pirro, D.M.; Daschner, E. *Lubrication Fundamentals*, 3rd ed.; Wessol, A.A., Ed.; CRC Press: Boca Raton, FL, USA, 2016.
70. Zum Gahr, K.-H. *Microstructure and Wear of Materials*; North Holland Publishing Co.: Amsterdam, The Netherlands, 1987.

MDPI
St. Alban-Anlage 66
4052 Basel
Switzerland
Tel. +41 61 683 77 34
Fax +41 61 302 89 18
www.mdpi.com

Coatings Editorial Office
E-mail: coatings@mdpi.com
www.mdpi.com/journal/coatings

www.ingramcontent.com/pod-product-compliance
Lightning Source LLC
LaVergne TN
LVHW070845170726
843515LV00005B/576

* 9 7 8 3 0 3 9 4 3 6 9 3 4 *